红色秘笈

巴尔塔沙·葛拉西安/著
秦传安/译
卜桦/插图

北京图书馆出版社

看上去很蠢的人都是蠢人，看上去不蠢的也多半是蠢人

THEY ARE ALL FOOLS WHO SEEM SO, AS WELL AS HALF THE REST.

蠢人与这个世界一起产生，即使还有点智慧的话，但跟神比起来也还是蠢人。不过最蠢的人，还是那位认为自己不蠢、别人都蠢的人。要做聪明人，看上去聪明是不够的，自认为聪明则更不行。不自以为知者是知也，不见人之所见者即不见也。尽管满世界都是蠢人，但没有一个人认为自己蠢，甚至怀疑这个事实。

Folly arose with the world, and if there be any wisdom it is folly compared with the divine. But the greatest fool is he who thinks he is not one and all others are. To be wise it is not enough to seem wise, least of all to seem so to oneself. He knows who does not think that he knows, and he does not see who does not see that others see. Though all the world is full of fools, there is no one who thinks himself one, or even suspects the fact.

快来避避雨！
笨蛋，伞都破了！

了解你同时代的伟人

KNOW THE GREAT PEOPLE OF YOUR AGE.

伟人并不多。全世界只有一只凤凰，一百年里也只有一位伟大的将领、一位完美的演说家、一位真正的哲学家，几百年才有一位出类拔萃的君王。平庸之辈比比皆是，毫无价值；各方面的杰出之士十分罕见，因为这需要十全十美，级别越高的方面越难达到它的至高点。许多人给自己冠以“伟大”的头衔，像恺撒和亚历山大那样，但无济于事，因为如果没有伟大的行为，头衔只不过是空穴来风。塞内加本来不多，青史留名的只有一个阿佩利斯。

They are not many. There is one phoenix in the whole world, one great general, one perfect orator, one true philosopher in a century, one really illustrious king in several. Mediocrities are as numerous as they are worthless; eminent greatness is rare in every respect, since it needs complete perfection, and the higher the species the more difficult is the highest rank in it. Many have claimed the title Great, like Caesar and Alexander, but in vain, for without deeds the title is a mere breath of air. There have been few Senecas, and fame records but one Apelles.

举重若轻，举轻若重

ATTEMPT EASY TASKS AS IF THEY WERE DIFFICULT AND DIFFICULT AS IF THEY WERE EASY.

举重若轻，可使精神不致沮丧；举轻若重，可使信心不致松懈。要想放弃一件未完之事不难，只需要认为它已经完成就行。从另一方面讲，锲而不舍可以征服不可能之事。对于宏图伟业，我们不能因为眼见得目标渺茫，唯恐其困难多多，就瞻前顾后，疑虑重重。

In one case so that confidence may not fall asleep, in the other so that it may not be dismayed. For a thing to remain undone nothing more is needed than to think it done. On the other hand, patient industry overcomes impossibilities. Great undertakings are not to be brooded over, lest their difficulty when seen causes despair.

www.buhua.com
2004

懂得如何打轻视这张牌

要得到你想要的东西，装作贬低它们是个精明的方法。通常，当你孜孜以求的时候总是得不到，而当你不再期望的时候，它们却会落入你的手中。既然尘世万物都只不过是永恒事物的影子，那它们也一样有影子的特点：你追它们，它们就逃离你；你逃离它们，它们就会追你。轻视也是最巧妙的报复形式。智者的铁律是：决不用手里的笔为自己辩护。因为这样的辩护总是留下污点，更多的是给对手增色，而不是惩罚他们的过错。跟伟人作对是卑鄙小人的惯用伎俩，这样他们就能够通过迂回曲折的方式赢得名声，这是他们通过直接的荣誉之路决不可能赢得的。有许多人，假如他们显赫的对手对他们置之不理的话，我们压根就不会知道他们。最好的报复就是遗忘，通过遗忘，他们被埋葬在卑贱的尘土里。轻妄之徒总是希望通过纵火烧掉某件世界和时代的奇迹而使自己名垂千古。谴责丑闻的艺术就是置之不理。与之争斗只能害及自己——即使是相信它也会让自己蒙羞，而让对手心满意足。这个污点所带来的阴影，让我们的名声黯然失色，即使不会让它完全失去光泽。

KNOW HOW TO PLAY THE CARD OF CONTEMPT.

It is a shrewd way of getting things you want, by pretending to depreciate them; generally they are not to be had when sought for, but fall into one's hands when one is not looking for them. As all mundane things are but shadows of the things eternal, they share with shadows this quality, they flee from him who follows them and follow him that flees from them. Contempt is also the most subtle form of revenge. It is a fixed rule with the wise never to defend themselves with the pen. For such defense always leaves a stain, and does more to glorify one's opponent than to punish his offence. It is a trick of the worthless to stand forth as opponents of great men, so as to win notoriety by a roundabout way, which they would never do by the straight road of merit. There are many we would not have heard of if their eminent opponents had not taken notice of them. There is no revenge like oblivion, through which they are buried in the dust of their unworthiness. An audacious person hopes to make himself eternally famous by setting fire to one of the wonders of the world and of the ages. The art of reproving scandal is not to take notice of it. To combat it damages our own case — even if credited it causes discredit and is a source of satisfaction to our opponent. This shadow of a stain dulls the luster of our fame, even if it cannot altogether deaden it.

看你能坚持多久！
BWhang 2007

要知道粗俗之人无处不在

KNOW THAT THERE ARE VULGAR PEOPLE EVERYWHERE.

即使是在科林斯①，即使是名门望族，也莫不如此。人人都可以在家门之内做个试验。但也有这样的事情：粗俗地反对粗俗，这更糟。这种特殊的粗俗，一样具有普通粗俗的所有特性，就像一片碎玻璃一样有玻璃的特性一样，但这种粗俗更加有害；他们说傻话，责备粗鲁者，他们是无知者的弟子，是笨蛋的保护人，是流言蜚语的行家里手。你不必留意他们说什么，更不要在乎他们想什么。重要的是要认识粗俗，为的是避开它，无论它是主观的还是客观的。所有愚蠢都是粗俗，粗俗之辈就是由蠢人所组成的。

This is true even in Corinth itself, even in the highest families. Everyone may try the experiment within his own gates. But there is also such a thing as vulgar opposition to vulgarity, which is worse. This special kind shares all the qualities of the common kind, just as bits of broken glass, but this kind is still more pernicious; it speaks folly, blames impertinently, is a disciple of ignorance, a patron of folly, a past master of scandal. You need not notice what it says, still less what it thinks. It is important to know vulgarity in order to avoid it, whether it is subjective or objective. For all folly is vulgarity, and the vulgar consist of fools.

---

①科林斯，一个以知识和文化而著称于世的古希腊城邦。

要有节制

一个人不得不考虑到灾难的可能性。激情的刺激让审慎在不知不觉间消失于无形，这就存在崩溃的危险。怒发冲冠或喜上心头的一瞬间，比起许许多多平静的时刻，能够把你带到更远，片刻的排遣可以让你终生蒙羞。他人的狡诈总是利用这样一些充满诱惑的瞬间，以探索你心灵的深处。他们用这样的刑具来检验你最好的警觉感。节制充当了一种对抗策略，尤其是在突如其来的紧急事件当中。要想像驾驭烈马那样驾驭激情，就需要深思熟虑，能在马背上审慎明智的人，是双倍的明智。认识到危险的人，就会小心翼翼地赶路。一言既出，在说者看来似乎微不足道，而对听到并仔细思量的人来说，却意义重大。

BE MODERATE.

One has to consider the chance of a mischance. The impulses of the passions cause prudence to slip, and there is risk of ruin. A moment of wrath or of pleasure carries you on farther than many hours of calm, and often a short diversion may put a whole life to shame. The cunning of others uses such moments of temptation to search the recesses of the mind. They use such thumbscrews to test your best sense of caution. Moderation serves as a counterplot, especially in sudden emergencies. Much thought is needed to prevent a passion taking the bit in the teeth, and he is doubly wise who is wise on horseback. He who knows the danger may with care pursue his journey. As light as a word may appear to him who throws it out, it may import much to him that hears it and ponders on it.

不要死于蠢病

智者通常死于丧失理智之后，蠢人总是死在找到理智之前。死于蠢病，就是死于太多的想法。有些人死去，是因为他们思考太多、感受太多；而有些人活着，则是因为他们压根就既不思考也不感受。前者是傻瓜，因为他们死于悲痛，后者则否。傻瓜就是死于知识太多的人。因此，有些人死是因为他们知道太多，而另一些人则是因为他们知道得不够。然而，尽管许多人像傻瓜一样死去，但死掉的傻瓜并不多。

DO NOT DIE OF THE FOOLS' DISEASE.

The wise generally die after they have lost their reason, fools before they have found it. To die of the fool's disease is to die of too much thought. Some die because they think and feel too much, others live because they do not think or feel at all. The first are fools because they die of sorrow, the others because they do not. A fool is he that dies of too much knowledge. Thus some die because they are too knowing, others because they are not knowing enough. Yet though many die like fools, few die fools.

自由

摆脱普遍的愚蠢

这是特殊的策略之举。普遍的愚蠢有着特殊的力量，因为它们是人所共有的，所以，许多不会被个人的愚蠢牵着鼻子走的人，却难逃人所共有的缺点。在这些普遍的愚蠢当中，应该算上对任何一个满足于自己的财富（无论多大）、不满于自己的智力（无论多弱）的人的普遍偏见。再或者，每个不满足自己的运气、嫉妒别人的运气的人。又或者，那些今日赞颂昨日之事、此处赞颂彼处之事的人。过去的每件事情似乎都是最好的，远方的每件事情都更加宝贵。嘲笑一切的人，跟哀哭一切的人一样蠢。

KEEP YOURSELF FREE FROM COMMON FOLLIES.

This is a special stroke of policy. They are of special power because they are common, so that many who would not be led away by an individual folly cannot escape the universal failing. Among these are to be counted the common prejudice of anyone who is satisfied with his fortune, however great, or unsatisfied with his intellect, however poor it is. Or again, that each, being discontented with his own lot, envies that of others. Or further, that persons of today praise the things of yesterday, and those here the things there. Everything past seems best and everything distant is more valued. He is a great fool that laughs at everything as is he that weeps at everything.

Bulnes 2007.

懂得如何打真相这张牌

真相很危险，然而好人就难免要说出真相。但这需要大技巧。最高明的心灵医生非常重视如何让真相的药丸变得更甜。因为当真相涉及到毁灭幻想的时候，它也就成了苦涩的精华。令人愉悦的方式，在这里就有了展示妙手的机会——同样的真相，既能让一个人高兴，也能让另一个人倒地。应该把今天的事情当作很久以前的事情来处理。对于那些能够理解的人来说，一句话就足够了，如果不够的话，就该保持沉默了。千万别用苦药来治疗君王，所以，在他们的病例中，最好是给幻灭的药丸裹上糖衣。

KNOW HOW TO PLAY THE CARD OF TRUTH.

It is dangerous, yet a good person cannot avoid speaking it. But great skill is needed here. The most expert doctors of the soul pay great attention to the means of sweetening the pills of truth. For when it deals with the destroying of illusion it is the quintessence of bitterness. A pleasant manner has here an opportunity for a display of skill — with the same truth it can flatter one and fell another to the ground. Matters of today should be treated as if they were long past. For those who can understand, a word is sufficient, and if it does not suffice, it is a case for silence. Princes must not be cured with bitter draughts, so it is desirable in their case to gild the pill of disillusion.

咋地?!
你不行啊?!
真正的交流都是精神上的!!

天堂与地狱

天堂里全是快乐，地狱中只有痛苦。红尘世俗，苦乐咸备。我们站在两极之间，端的是苦乐皆尝。风水轮回，祸福莫测：既不全是幸运，亦非尽是倒霉。小小寰球，只不过是个零，孤立地看，它毫无价值，但是有个天堂在它前面，这就非同小可。管它世事浮沉，我自宠辱不惊，方是明智之策，聪明人甚至对它的任何新鲜神奇也毫不上心。我们的生活就像一出正在进行的喜剧一样复杂难解，不过这样的复杂性也在逐步解开——当一个美好结局来临的时候，大幕也就徐徐落下。

IN HEAVEN ALL IS BLISS.

And in hell all misery. One earth, between the two, both one thing and the other. We stand between the two extremes, and therefore share both. Fate varies — all is not good luck nor all mischance. This world is merely zero — by itself it is of no value — but with heaven in front of it, it means much. Indifference at its ups and downs is prudent, nor is there any novelty for the wise. Our life gets as complicated as a comedy as it goes on, but the complications get gradually resolved - see that the curtain comes down on a good denouement.

www.buhua.com
2004

把最后几招留给自己

这是大师的座右铭，他们在教弟子时就是采用这种精明的策略，并引以为傲。你必须始终保持技高一筹，始终是师傅。你必须巧妙地传授技艺。知识的源头大可不必指出，就像礼物的来源一样。借助这种手段，你就可以维持他人对你的尊敬和依赖。在博人一笑和授人以技的时候，你必须遵循这样的准则：要维持他人对你的期待以及留待完美的余地。凡事留一手，是生活和成功的伟大准则，身处高位者，尤应如此。

KEEP TO YOURSELF THE FINAL TOUCHES OF YOUR ART.

This is a maxim of the great masters who pride themselves on this subtlety in teaching their pupils: one must always remain superior, remain master. One must teach an art artfully. The source of knowledge need not be pointed out no more than that of giving. By this means a person preserves the respect and the dependence of others. In amusing and teaching, you must observe the rule: keep up expectation and advance in perfection. To keep a reserve is a great rule for life and for success, especially for those in high places.

懂得如何驳难

这是揭示事物真相的主要手段——置他人于窘境而自己不为所窘。这是真正的夹指刑具，它让激情得以展现。稍加质疑，就让人吐露秘密。它是打开上锁心灵的钥匙，大智大巧给头脑和意志制造了双重的考验。机智巧妙的贬损，能从他人故弄玄虚的言词中嗅出深邃的奥秘；少许香甜的诱饵，能把这些秘密从心底带至口端，直至从舌尖滑落，被精明的谎言所编织的网所捕获。保留你的注意，别人就变得不大注意，在他的心莫测高深的时候，就让他的想法浮现出来。假装的怀疑，是巧妙的撬锁工具，好奇心可以用它来找出自己想知道的东西。在学问上也是如此，驳难老师是弟子的妙计，这样一来，老师就会费尽心机地花更大的力气把真理解释得更透彻。因此，适度的反驳，总是带来更完善的教导。

KNOW HOW TO CONTRADICT.

A chief means of finding thing out — to embarrass others without being embarrassed. The true thumbscrew, it brings the passions into play. A little disbelief makes people spit up secrets. It is the key to a locked up heart, and with great subtlety makes a double trial of both mind and will. A sly depreciation of another's mysterious word scents out the profoundest secrets; some sweet bait brings them into the mouth till they fall from the tongue and are caught in the net of astute deceit. By reserving your attention the other becomes less attentive, and lets his thoughts appear while otherwise his heart were inscrutable. An affected doubt is the subtlest picklock that curiosity can use to find out what it wants to know. Also in learning it is a subtle plan of the pupil to contradict the master, who thereupon takes pains to explain the truth more thoroughly and with more force, so that a moderate contradiction produces complete instruction.

你不可能是女的！
不信给你看！

不要一错再错

为纠正一件错事而再犯四次错，或者用一次更大的失当去开脱另一次失当，这实在太稀松平常了。蠢行与谎言，要么是亲戚，要么是一家子。一件蠢行需要更多的蠢行去支撑，谎言也是如此。一件坏事的最坏之处，是你不得不与它战斗；比坏本身更坏的，是你没法掩盖它。一次失败带来的年金，可以供养多次失败。智者也难免摔一次跟斗，但决不会摔第二次，而且，他只会在奔跑中跌翻，而不是在站着的时候滑倒。

DO NOT TURN ONE BLUNDER INTO TWO.

It is quite usual to commit four blunders in order to remedy one, or to excuse one piece of impertinence by still another. Folly is either related to or identical with the family of lies, for in both cases it needs many to support one. The worst of a bad case is having to fight it, and worse than the ill itself is not being able to conceal it. The annuity of one failing serves to support many others. A wise person may make one slip but never two, and that only in running not while standing still.

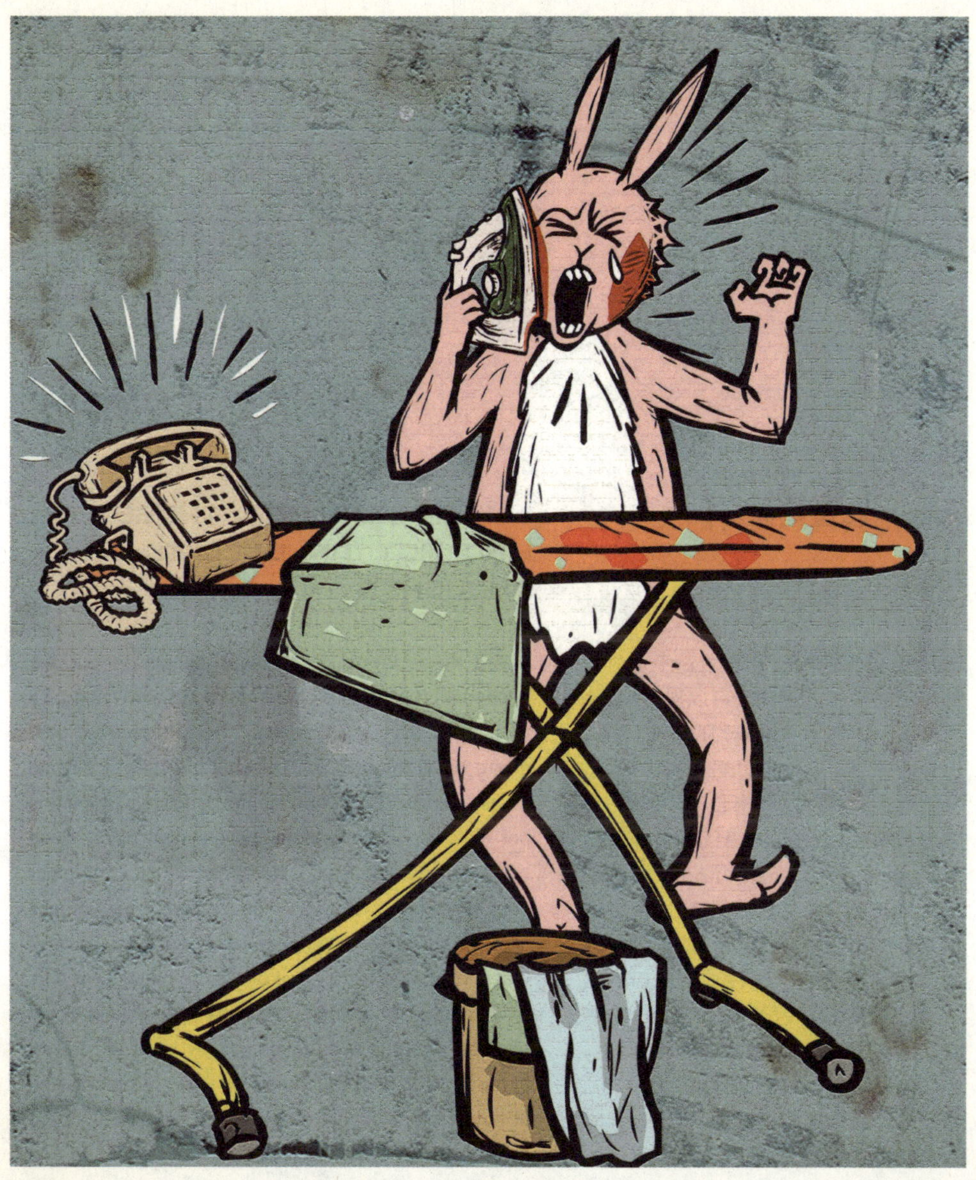

提防那些再思而行的人

商人的谋略是:在攻击对手之前先卸掉他的防备，这样就可以挫败他,从而征服他。他们总是掩饰自己的愿望,以便能得到它。他们甘居其次,为的是独占鳌头。这种伎俩,倘若不加留意,常常屡试不爽。因此,当意图睁大眼睛的时候,就千万别让注意力酣睡。如果别人甘居其次以便隐藏他的计划,那么你就当仁不让,为的是发现他的图谋。遇事谨慎，便能洞悉这种人所使用的花招，注意到他为达到目的而提出的借口。他为了得到彼物而瞄准此物,然后,他潇洒地转身，直接向他的靶子射击。最好是知道你能给予他的是什么,有时候,让他了解你所了解的也是值得的。

WATCH OUT FOR THOSE WHO ACT ON SECOND THOUGHTS.

It is a device of business people to put the opponent off his guard before attacking him, and thus to conquer by being defeated. They dissemble their desire so as to attain it. They put themselves second so as to come out first. This method rarely fails if it is not noticed. Let therefore the attention never sleep when the intention is so wide awake. And if the other puts himself second so to hide his plan, put yourself first to discover it. Prudence can discern the artifices that such a man uses, and notice the pretexts he puts forward to gain his ends. He aims at one thing to get another, then he turns round smartly and fires straight at his target. It is good to know what you grant him, and at times it is desirable to let him understand that you understand.

要善于表达

这不仅仅取决于表达的清晰，也取决于思考的活泼。有些人怀孕很轻松，分娩却很难，因为，如果没有清晰的表达，头脑的产儿——思考和判断——就不可能生下来。许多人颇有容量，就像肚大口小的容器。而有的人却说的比想的多。决心为意志而生，表达为思考所用——二者都是天赋。似是而非的想法受到人们的喝彩，混乱不堪的思路常常受到尊崇，这恰恰是因为人们没有搞懂——有时候，如果你希望避免粗言俚语的话，含糊其辞倒是个省力的法子。一个人的明确想法要是跟他所说的话毫无关系，听众又如何能懂呢？

BE EXPRESSIVE.

This depends not only on the clearness but also on the vivacity of your thoughts. Some have an easy conception but a hard labor, for without clarity the children of the mind — thoughts and judgements — cannot be brought into the world. Many have a capacity like that of vessels with a large mouth and a small vent. Others say more than they think. Resolution for the will, expression for the thought — both are gifts. Plausible minds are applauded, yet confused ones are often venerated just because they are not understood — at times obscurity is convenient if you wish to avoid vulgarity. How will the audience understand someone who does not connect and definite idea with what he is talking about?

既没有永恒的爱，也没有永恒的恨

对今天的朋友付出信任时，要想到明天他们会成为敌人（而且是最坏的那种）。既然这种事情会在现实中发生，那就让它发生在你有所防范的时候。不要把武器放到友谊的背叛者手里，那样他们就会用这些武器来发动战争。从另一方面讲，要为敌人留下敞开的和解之门，而且，如果它同时又是慷慨之门的话，那就更加安全。很久之前的复仇，有时是今天的折磨；我们曾经因作恶所带来的快乐，有时也会转化为内心的伤痛。

NEITHER LOVE NOR HATE FOREVER.

Trust the friends of today as if they will be enemies tomorrow, and that of the worst kind. As this happens in reality, let it happen in your precaution. Do not put weapons in the hand for deserters from friendship to wage war with. On the other hand, leave the door of reconciliation open for enemies, and if it is also the gate of generosity so much the more safe. The vengeance of long ago is at times the torment of today, and the joy over the ill we have done is turned to grief.

www.buhua.com
2004

行事不要出于固执，而要出于知识

NEVER ACT FROM OBSTINACY BUT FROM KNOWLEDGE.

一切固执都是头脑中的毒瘤，是激情的子孙后代，而激情，从不做什么正确的事。有的人事事大动干戈，是社会交往中真正的强盗。他们所着手的一切，都必须以胜利告终。他们不懂得如何和平共处。这样的人要是统治国家的话，那将是毁灭性的，因为他们会把政府搞成一场造反，而把那些本该被他们视为子女的人视为仇敌。对每件事情，他们都试图用兵法去实现，并视之为自己的技巧所带来的成果。但是，当其他人认识到了他们刚愎自用的脾性的时候，就会反抗他们，学着颠覆他们异想天开的计划。除了积聚起一大堆的麻烦之外，他们什么事也干不成，因为每件事情都有助于增加他们的失望。他们有一颗被扭曲的脑袋和一颗被宠坏的心。跟这样的怪物在一起，你所能做的只有逃得远远的——哪怕是野蛮人的野性也比他们令人厌恶的本性更容易忍受些。

All obstinacy is an evil tumor on the mind, a grandchild of passion that never did anything right. There are people who make a war out of everything, real bandits of social intercourse. All that they undertake must end in victory. They do not know how to get on in peace. Such people are fatal when they rule and govern, for they make government a rebellion and enemies out of those they should regard as children. They try to effect everything with strategy and treat it as the fruit of their skill. But when others have recognized their perverse humor, they revolt against them and learn to overturn their chimerical plans. They succeed in nothing but only heap up a mass of troubles, since everything serves to increase their disappointment. They have a head turned and a heart spoilt. Nothing can be done with such monsters except to flee from them — even the savagery of barbarians is easier to bear than their loathsome nature.

不要被人看作是伪君子

尽管现如今这样的人不可或缺。与其被人视为精明,不如被人视为审慎。诚实不该退化为简单,而睿智也不该堕落为狡诈。因为聪明而受人尊敬,总比由于狡猾而被人害怕要好。胸无城府者被人喜爱,但也常常被人欺骗。大智大巧就在于揭示什么东西会被认为是欺骗。简单流行于黄金时代,狡诈兴盛于冷铁岁月。知道自己该做什么的人,他的声誉值得尊敬,而且激发人们对他的信任;但如果被认为是个伪君子的话,那他的名望就是欺骗性的,并且引起人们对他的猜疑。

DO NOT PASS FOR A HYPOCRITE.

Though nowadays, such people are indispensable. Be considered prudent rather than astute. Sincerity should not degenerate into simplicity nor sagacity into cunning. Be respected as wise rather than feared as sly. The openhearted are loved but often deceived. The great art consists in disclosing what is thought to be deceit. Simplicity flourished in the golden age, cunning in these days of iron. The reputation of someone who knows what he has to do is honorable and inspires confidence, but to be considered a hypocrite is deceptive and arouses mistrust.

可不能跟这孙子过事儿！

披不了狮皮，你就披狐皮

紧跟时代，为的是引领时代。得到自己想要的东西，你决不会丧失名誉。力之不逮，智以取之。或由此路，或经彼途；或走勇敢的王者大道，或取狡诈的曲折小径。虽殊途同归，智尤胜于力。谋胜勇者甚多，勇胜谋者盖寡。如果你得不到那个东西，那就鄙视它。

IF YOU CANNOT CLOTHE YOURSELF IN LION-SKIN USE FOXPELT.

To follow the times is to lead them. He that gets what he wants never loses his reputation. Use cleverness when force will not do. Take one way or another, the king's highway of valor or bypath of cunning. Skill has effected more than force, and astuteness has conquered courage more often than the other way around. When you cannot get something, that is the time to despise it.

www.buhua.com
2004

别抓住机会让自己或他人难堪

DO NOT SEIZE OCCASIONS TO EMBARRASS YOURSELF OR OTHERS.

有些人总是摔倒在言行举止的绊脚石上，要么是因为他们自己，要么是因为别人。他们总是处于愚蠢的边缘。遇上他们很容易，摆脱他们却很难。一天惹上一百桩麻烦对他们来说根本不算一回事。他们的心境总是以错误的方式表现出来，因为他们跟所有人作对，跟所有事情作对。他们把判断的帽子反着戴，因此跟所有人对着干。然而，对他人的耐心和审慎来说，最大的考验恰恰是那些什么事情都做不好却对所有事情吹毛求疵的人。在粗俗无礼的广阔领域里，总是有许多这样的怪物。

There are some people who are stumbling blocks of good manners either for themselves or for others. They are always on the verge of some stupidity. You meet with them easily and part from them uneasily. A hundred annoyance a day is nothing to them. Their humor always strokes the wrong way since they contradict all and everything. They put on the judgement cap backwards and thus condemn all. Yet the greatest test of others' patience and prudence are just those who do no good and speak ill of all. There are many monsters in the wide realm of indecorum.

矜持是审慎的明证

舌头就是一头野兽——一旦放了出去，就很难再把它关进牢笼。它是灵魂的脉搏，智者可以根据它来判断一个人是否健康。通过这种脉搏，一个细心的观察者能感觉到心灵的每一次跳动。最糟糕的是，那些本该最缄默的人话却最多。圣贤总是避免陷入烦恼和困窘，显得能够控制自己。他小心翼翼地走自己的路，他是不偏不倚的罗马门神杰纳斯，是时刻警觉的百眼巨人阿耳戈斯。吹毛求疵的嘲弄之神莫摩斯，肯定更愿意把眼睛放在手上，而不是把窗户开在胸膛①。

① 在古希腊作家卢西恩所写的一篇故事中，莫摩斯嘲笑赫斐斯托斯在造人的时候没有在他的胸膛上开一扇窗户。

RESERVE IS PROOF OF PRUDENCE.

The tongue is a wild beast — once let loose it is difficult to chain. It is the pulse of the soul by which wise men judge its health. By this pulse a careful observer feels every movement of the heart. The worst is that he who should be most reserved is the least. The sage saves himself from worries and embarrassments, and shows his mastery over himself. He goes his way carefully, a Janus of impartiality, an Argus of watchfulness. Certainly Momus would have better placed the eyes in the hand than the windows in the breast.

不要成为怪人

DO NOT BE ECCENTRIC, NEITHER FROM AFFECTATION NOR CARELESSNESS.

行为古怪者，要么是因为装模作样，要么是由于粗心大意。许多人因为有某些非凡而独特的才能，从而导致了古怪的行为。这与其说是卓异，不如说是缺点。正如有些人因为相貌奇丑而广为人知，行为古怪者也正是因为他们某些令人厌恶的外在举止而臭名昭彰。像这样的怪癖只不过是通过他们糟糕透顶的特异而成为某种商标——其所能招来的，要么是嘲弄，要么是恶意。

Many have some remarkable and individual quality leading to eccentric actions. These are more defects than excellent differences. And just as some are known for some special ugliness, so these for something repellant in their outward behavior. Such eccentricities simply serve as trademarks through their atrocious singularity — they cause either derision or ill will.

www.buhua.com
2004

不要盯住事物粗糙的一面，无论它们出现的时候是怎样

NEVER TAKE THINGS AGAINST THE GRAIN, NO MATTER HOW THEY COME.

任何事物都有它光鲜的一面，也有粗糙的一面。最好的武器，如果抓住刃口也会伤着你；而即便是敌人的长矛，如果抓住矛杆，也是自己最好的保护。任何事情，都有有利的一面，也有不利的一面——聪明的做法，就在于找出有利的。同样的事情，如果站在不同的角度，看上去完全不同；因此要从最好的一面去看待事物，而不要善恶颠倒。这就是为什么许多人能从每件事情中发现快乐，而另一些人总是发现悲伤。对于时运不济，这种论点是一个很好的保护，无论任何时期，不管什么境遇，它都是生活的重要法则。

Everything has a smooth and a seamy side. The best of weapons wounds if taken the blade, while the enemy's spear may be our best protection if taken by the staff. Many things cause pain that would cause pleasure if you regarded their advantages. There is a favorable and an unfavorable side to everything — cleverness consists in finding out the favorable. The same thing looks quite different in another light; look at it therefore on its best side and do not exchange good for evil. Thus it happens that many find joy, many grief, in everything. This remark is a great protection against the frowns of fortune, and a weighty rule of life for all times and all conditions.

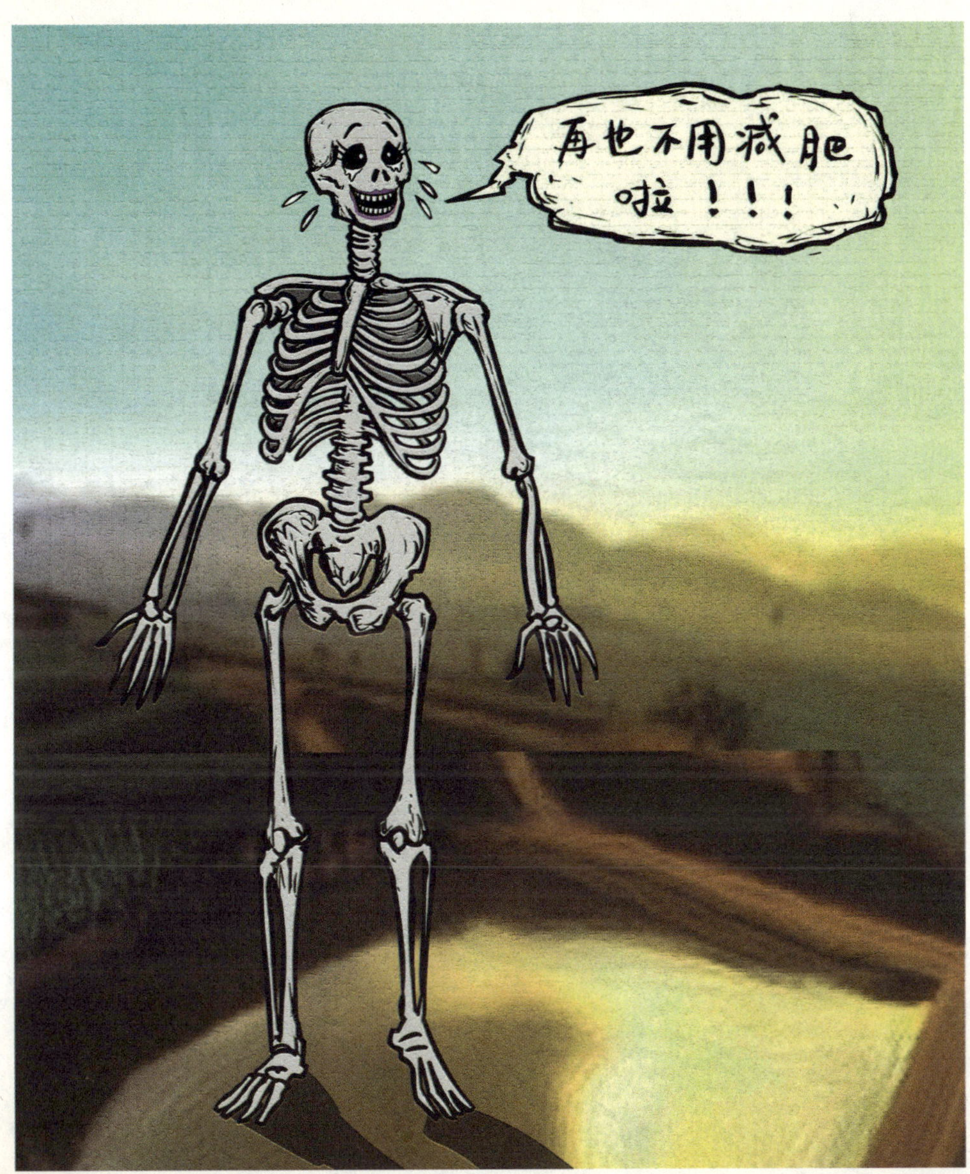
再也不用减肥啦！！！

了解自己的主要缺点

无一例外，每个人身上都有一种与自己显著优点相抗衡的东西，如果任由其滋长，它就有可能成长为一个暴君。丌始和它作战吧，召集起“审慎”作为你的盟友。你要做的第一件事，就是把它揪出来，暴露在公众面前，因为邪恶一旦被认识，很快就会被征服，尤其，当你以和旁观者同样的观点痛苦地关注它的时候，更是如此。要想成为自己的主人，你就应该了解自己。如果主要缺点举手投降，其余的也就会束手就缚。

KNOW YOUR CHIEF FAULT.

There is no one lives who has not in himself a counterbalance to his most conspicuous merit, and if it is nourished by desire it may grow to be a tyrant. Commence war against it, summoning prudence as your ally. The first thing to do is to make it public, for an evil once known is soon conquered, especially when the one afflicted regards it in the same light as the onlookers. To be master of oneself one should know oneself. If the chief imperfection is surrendered, the rest will also come to an end.

www.buhua.com
2004
Z Z Z
Z Z Z

留心让别人感到满意

多数人的言行并非出于本性，而是出于无奈。任何人，要想说服人们相信坏事并不困难，因为坏事容易让人相信，哪怕它确实难以置信。我们所拥有的最好的一面，也依赖于他人的看法。有些人只要正义在自己这边，就感到心满意足，但这远远不够，我们还必须拿出干劲助其一臂之力。使别人感到满意，付出甚少，获益良多。你用言辞买到了行动。在世界这个大宅子里，没有哪一间房子会如此隐秘，以至于一年也没人需要它一次，而且，不管它多么微不足道，你会因而缺少它而痛苦。每个人都根据自己的感觉谈论某个话题。

TAKE CARE TO BE OBLIGING.

Most talk, and act, not as they are, but as they are obliged. To persuade people of the bad is easy for anyone, since the bad is easily credited even when it is incredible. The best we have depends on the opinion of others. Some are satisfied if they have right on their side, but that is not enough, for it must be assisted by energy. To oblige people often costs little and helps much. With words you may purchase deeds. In this great house of the world there is no chamber so hidden that it may not be wanted one day in the year, and then you would miss it however little it is worth. Everyone speaks of a subject according to his feelings.

www.buhua.com
2004

不要做第一印象的奴隶

有些人把他们最初的道听途说娶为正室,一切后来的都只能是偏房。但是,因为谎言有一双飞毛快腿,尾随其后的真理因而总是找不到落脚之处。我们的意愿既不能满足于最初的目标,我们的头脑也不能满足于最先的命题——因为那是肤浅的。许多人就像崭新的木桶,总是保持着其装盛的第一种液体的气味,也不管它是好是坏。如果这样的浅薄广为人知,它也就成了致命的缺点,因为这之后他就给狡猾的伤害留出了机会。那些用心险恶的人就会赶紧把这颗容易上当受骗的头脑染成他们希望的颜色。因此,始终为你第二次听到的东西留出容身之地吧。亚历山大总是为另一面留出一只耳朵。对于一则消息,要等待它的第二个甚至第三个版本。成为第一印象的奴隶,就显得你缺乏度量,而且,这离成为激情的奴隶,实在也相去不远。

DO NOT BE THE SLAVE OF FIRST IMPRESSIONS.

Some marry the very first account they hear, all others must live with them as concubines. But as a lie has swift legs, the truth with them can find no lodging. We should neither satisfy our will with the first object nor our mind with the first proposition — for that is superficial. Many are like new casks who keep the scent of the first liquor they hold, be it good or bad. If this superficiality become known, it becomes fatal, for it then gives opportunity for cunning mischief. The evil-minded hasten to color the mind of the gullible. Always therefore leave room for a second hearing. Alexander always kept one ear for the other side. Wait for the second or even third edition of news. To be the slave of your first impressions shows lack of capacity, and is not far from being the slave of your passions.

www.buhua.com
2004

不要做传播流言蜚语的人

更不要被人误认为是这样的人，因为那意味着你将被视为一个造谣中伤者。不要以损害别人为代价来表现自己的诙谐机智，这样做并不难，但却招人憎恨。对这样一个人，每个人都会报复，说他的坏话，他们人多势众，而你只有独自一人，你更大的可能是被人击败，而不是令人信服。邪恶绝不是我们的乐趣之所在，因此也绝不能作为我们的话题。背后说人坏话的人总是被人憎恨，即使偶尔跻身伟大人物的行列，人们从他的冷嘲热讽中所得到的乐趣，也比从对其洞察力的尊敬中所得到的乐趣要少。说人坏话的人总是听到更坏的话。

DO NOT BE A SCANDALMONGER.

Still less pass for one, for that means to be considered a slanderer. Do not be witty at the cost of others; it is easy but hateful. Everyone will have their revenge on such a person by speaking ill of him and, as they are so many and he but one, he is more likely to be overcome than they convinced. Evil should never be our pleasure and therefore never our theme. The backbiter is always hated, and if now and then one of the great consorts with him it is less from pleasure in his sneers than from esteem for his insight. He that speaks ill will always hear worse.

明智地规划你的一生

不要听天由命，而是要审慎而富有远见地规划你的一生。没有娱乐，生活就很乏味，就像是经过漫长的跋涉，来到一个没有旅馆的地方一样——多方面的知识会带来多方面的乐趣。高贵的一生，第一天的旅行应该在与死者的交谈中度过：我们活着就是要去认识外部世界，认识自己，因此，真正的书籍会让我们真正地长大成人。第二天应该跟生者一起度过：留心观察世间一切美好的事物。在一个孤家寡人的国度里，任何事情也发现不了。宇宙这位父亲总是把他的礼物分成很多份，有时候把最值钱的嫁妆给他最难看的女儿。第三天应该完全独自一人。最大的幸福莫过于成为一位哲人。

PLAN OUT YOUR LIFE WISELY.

Not as chance would have it, but with prudence and foresight. Without amusements it is wearisome, like a long journey where there are no inns — manifold knowledge gives manifold pleasure. The first day's journey of a noble life should be passed in conversing with the dead: we live to know and to know ourselves, hence true books make us truly human. The second day should be spent with the living, seeing and noticing all the good in the world. Everything is not to be found in a single country. The Universal Father has divided his gifts and at times has given the richest dowry to the ugliest. The third day is entirely for oneself. The greatest happiness is to be a philosopher.

及早睁开你的眼睛

并非所有能看的人都睁开了眼，能看的人也并非人人都有所见。认识事物太迟，更多的是烦恼，而非助益。有些人只有到了没什么东西可看的时候才开始去看：没等他们醒悟过来，他们就把自己的房子从头推倒。要把理解力赋予那些没有意志力的人固然很难，而要把意志力赋予那些没有理解力的人则难上加难。周围的人跟他们玩捉迷藏的游戏，让他们成为笑柄。因为他们耳朵不好使，又不睁开眼睛看。常常有人怂恿这样的无知无识，因为他们的生存就靠这个。遇上盲骑手的战马是不幸的，它绝不会长得膘肥体壮。

OPEN YOUR EYES EARLY.

Not all who see have their eyes open, nor do all those see who look. To come up on things too late is more worry than help. Some just begin to see when there is nothing more to see: they pull their houses down about their heads before they come to themselves. It is difficult to give understanding to those who have no power of will, still more difficult to give power of will to those who have no understanding. Those who surround them play a game of blindman's buff with them, making them the butts of jokes. Because they are hard of hearing, they do not open their eyes to see. There are often those who encourage such insensibility because their very existence depends on it. It is an unhappy steed whose rider is blind: it will never grow sleek.

未完成的事情别让人看见

事情只有在完成之后才能被人们欣赏。所有事情一开始都是残缺不全的，这样的畸形固定在人们的想象中。当一件事情完成的时候，回想起当初看到的残缺，就会妨碍我们对它的欣赏。一口吞下某件巨物，或许会妨碍我们对各局部的判断，但口味却更好。一件事情在“有”之前就是“无”，虽然它正在“有”的过程中，但依然是“无”。看见装盛美味的盘碟，非但不会引起食欲，相反还会倒人胃口。每一位大师都应该多加留意，在你的作品还处于孕育生成阶段的时候，不要让它被人看见——他们其实可以从大自然中学习这一课：大自然从不会让自己的孩子轻易露面，除非它已经适合被人看见。

NEVER LET THINGS BE SEEN HALF FINISHED.

They can only be enjoyed when complete. All beginnings are misshapen, and this deformity sticks in the imagination. The recollection of having seen a thing imperfect disturbs our enjoyment of it when completed. To swallow something great at one gulp may disturb the judgement of the separate parts, but satisfies the taste. Before a thing is manifest, it is nothing, and while it is in process of being it is still nothing. To see the tastiest dishes prepared arouses disgust rather than appetite. Let each great master take care not to let his work be seen in its embryonic stages — they might take this lesson from Mother Nature, who never brings the child to the light till it is fit to be seen.

www.buhua.com
2004

干点实事

生活不应该全都是思考，还应该有行动。非常聪明的人通常容易上当受骗，因为他们虽说熟知那些非同寻常的事物，却对生活中稀松平常的事物一无所知，这些才是真正必不可少的。观察更高级的事物，使得他们没有时间来思考近在眼前的事物。既然他们对自己本该知道、而且人人都知道的首要事物一无所知，那么，他们要么被见识肤浅的大众所敬重，要么被视为无知。因此，审慎之人应该操心一点实干家的事——这足以防止他上当受骗，被人嘲弄。做一个适应日常事务的人吧，这在生活中即使不是最高级的，但肯定是最必不可少的。知识如果不能实践，那要它何用？现如今，懂得如何生活就是真知识。

HAVE A TOUCH OF BUSINESS SENSE.

Life should not be all thought, there should be action as well. Very wise folk are generally easily deceived, for while they know out-of-the-way things they do not know the ordinary things of life, which are of real necessity. The observation of higher things leaves them no time for things close at hand. Since they do not know the very first thing they should know — and what everybody knows so well — they are either esteemed or thought ignorant by the superficial multitude. Let therefore the prudent take care to have something to the businessman about him — enough to prevent him being deceived and so laughed at. Be a person adapted to the daily round, which if not the highest is the most necessary thing in life. Of what use is knowledge if it is not practical, and to know how to live is nowadays the true knowledge.

别让你奉献的佳肴不合别人的口味

否则的话，他们给你的更多的是难堪，而不是愉快。有些人在努力取悦别人的时候总是让别人生气，因为他们没有考虑品味的不同。有的事，对某个人是奉承，而对另一个人却是冒犯，你好心帮忙，到头来却成了侮辱。冒犯某个人，其代价常常比取悦他的代价更高——你将因此既丢掉礼物又丢掉感谢，因为你已经丢掉了驾驭愉悦的罗盘。因此，许多人总是在他们本想大唱赞歌的时候给人带来侮辱，结果受到严厉的惩罚，这是罪有应得。另一些人则希望通过他们的谈话来吸引别人，结果只成功地用他们的喋喋不休让别人不胜其烦。

DO NOT LET THE MORSELS YOU OFFER BE DISTASTEFUL.

Otherwise they give more discomfort than pleasure. Some annoy when attempting to please, because they take no account of varieties of taste. What is flattery to one is offense to another, and in attempting to be useful you may become insulting. It often costs more to displease someone than it would have cost to please him — you thereby lose both gift and thanks because you have lost the compass that steers for pleasure. If you do not know another's taste, you do not know how to please him. Thus it happens that many insult where they mean to praise, and get soundly punished, and rightly so. Others desire to charm by their conversation, and only succeed in boring by their babble.

哇！你比猪漂亮多啦！

不要把你的名誉托付给他人，除非拿他的名誉作担保

NEVER TRUST YOUR HONOR TO ANOTHER, UNLESS YOU HAVE HIS IN PLEDGE.

对于双方来说，有了这样的安排，沉默则互惠，暴露则俱危。既然名誉得失攸关，行动就必须彼此协调。这样一来，每个人为了自己的利益，就必定会小心对待另一方的名誉。不要把你的名誉完全托付给他人，如果你不得不这样做，与其小心谨慎，不如警告在先。让威胁共有，而风险同担，这样，你的伙伴就不会“出庭提供同案犯的罪证”。

Arrange that silence is a mutual advantage, disclosure a danger to both. Where honor is at stake you must act with a partner, so that each must be careful of the other's honor for the sake of his own. Never fully entrust your honor to another, but if you have to, let caution surpass prudence. Let the danger be in common and the risk mutual, so that your partner cannot turn king's evidence.

www.buhua.com
2004

懂得怎样求人

求人一事，对于有些人来说，最易，而在另一些人而言，最难。有人从不知如何拒绝别人——对付这样的人无须技巧。有人却总是张口说“不”——对付这样的人则大费周章，而且还要选择合适的时机。赶在他们心情愉快的时候，在他们身体和精神都饱餐一顿之后而心神俱爽的时候，给他们一个措手不及——不过，也只有在他们的精明尚没有预先觉察到你的小把戏的时候，才能这样。欢喜愉快的日子也是乐于助人的日子，因为欢乐从身体内部漫溢而出，就会流向外部世界。在别人刚刚遭到拒绝的时候，你再去请求将无功而返，因为“不”字的难以出口已经被他克服了。悲痛过后也不是一个好时机。预先让一个人对你心怀感激是一个可靠的办法，除非他是一个卑劣吝啬之徒。

KNOW HOW TO ASK.

With some nothing is easier, with others nothing is so difficult. For there are men who cannot refuse — with them no skill is required. But with others their first word at all times is no — with them great art is required, and with everyone pick the right moment. Surprise them when they are in a pleasant mood, when a repast of body or soul has just left them refreshed — but only if their shrewdness has not anticipated your cunning. The days of joy are the days of favor, for joy overflows from the inner person into the outward creation. It is no use to apply when another has just been refused, since the reticence of saying no has just been overcome. Nor is it good time after sorrow. To oblige a person beforehand is a sure way, unless he is base and mean.

www.buhua.com
2004
YES
NO

把后来应该成为报酬的东西预先作为恩惠施与出去

MAKE AN OBLIGATION BEFOREHAND OF WHAT WOULD HAVE TO BE A REWARD AFTERWARD.

这是一招精明的策略。在恩惠成为理所当受之前把它们施与出去，这是乐于助人的明证。这样预先施与的恩惠，有两大好处：礼物的提前出手，会让接受者更加感激；而且，同样的礼物，后出手只是报酬，先出手则是恩惠。这是把报酬转变为恩惠的一种微妙手段，因为，迫使你不得不酬赏某个人的东西，被改变成了迫使他们不得不偿还的人情债。但这只适用于那些知道感恩的人，因为对更卑劣的人来说，预先支付的报酬是一种控制，而非鞭策。

This is a stroke of subtle policy. To grant favors before they are deserved is a proof of being obliging. Favors thus granted beforehand have two great advantages: the promptness of the gift obliges the recipient more strongly. And the same gift that would afterward be merely a reward is beforehand an obligation. This is a subtle means of transforming obligations, since that which would force you to reward someone is changed into one that obliges them to satisfy their obligation. But this is only suitable for people who feel obligation, since with people of lower stamp the honorarium paid beforehand acts rather as a bit than as a spur.

礼物!!
晚饭

不要和上司分享秘密

NEVER SHARE THE SECRETS OF YOUR SUPERIORS.

你或许认为自己会分到梨子，结果却只能分到削下来的皮。许多人因为成为别人的心腹知己而毁了自己：他们就像用面包片做成的汤匙，同样冒着随后被吞掉的危险。分享一位君王的秘密并不是恩宠——你只不过是在分享他的负担。许多人打碎镜子，是因为镜子让他们想起自己的丑陋。我们不愿意见到那些曾经看见过我们本来面目的人，而你如果看见过别人不利的一面，别人也就不会带着嘉许的目光看待你。没有谁应该对你感恩戴德(尤其是大人物)，除非你给过他帮助，而不是接受了他的帮助。特别危险的是把秘密托付给朋友。当你把秘密传达给他人的时候，你就使自己成为他的奴隶。处于这样的位置是君王所不能忍受的，因此也就不能持久；他会千方百计要恢复失去的自由，为此不惜颠覆一切，包括正义和理性。因此，秘密就是秘密，既不能说，也不要听。

You may think you will share pears, but you will only share parings. Many have been ruined by being confidants: they are like sops of bread used like spoons, they run the same risk of being eaten up afterwards. It is no favor to a prince to share a secret — it is only a relief. Many break the mirror that reminds them of their ugliness. We do not like seeing those who have seen us as we are, nor is he seen in a favorable light who has seen us in an unfavorable one. No one ought to be too much beholden to us, least of all one of the great, unless it is for favors done for him rather than for favors received. Especially dangerous are secrets entrusted to friends. When you communicate a secret to someone you make yourself his slave. With a prince this is an intolerable position that cannot last; he will desire to recover his lost liberty, and to gain he will overturn everything, including right and reason. Accordingly, neither tell secrets nor listen to them.

www.buhua.com
2004

知道自己缺少什么

许多人如果不是缺少点什么的话，应该会成为了不起的人物，有了这些他们才会达到完美的高度。值得注意的是，有些人只要在某些方面稍微好一点点，他们就会好很多。他们多半没有足够认真地对待自己，以至于对自己的能力不能做出公正的判断。有些人缺乏温和的脾气，这正是身边的人希望从他们身上找到的一种品质，位高权重者尤其如此。有的人没有组织能力，有的人则缺乏克制力。在所有这样的情况下，谨慎细心的人就会让习惯成为自己的第二天性。

KNOW WHAT IS LACKING IN YOURSELF.

Many would have been great people if they had not had something wanting, without which they could not rise to the height of perfection. It is remarkable that some people could be much better if they could be just a little better in something. They do not perhaps take themselves seriously enough to do justice to their great abilities. Some are lacking geniality of disposition, a quality which their entourage soon finds want of, especially if they are in high office. Some are without organizing ability, others lack moderation. In all such cases a careful person may make of habit a second nature.

不要过于挑剔

更重要的是要通情达理。知道得太多会让你的武器变钝，因为锐利的尖端通常容易弯曲，或者断裂。常识的真理是最可靠的。最好是去了解，而不是去挑剔。冗长的评论总是导致争辩。要是有健全的判断力则要好得多，它不会离手边的事情太远。

DO NOT BE OVERLY CRITICAL.

It is much more important to be sensible. To know more than is necessary blunts your weapons, for fine points generally bend or break. Commonsense truth is the surest. It is well to know but not to niggle. Lengthy comment leads to disputes. It is much better to have sound sense, which does not wander from the matter in hand.

没骨头！！

利用愚蠢

最聪明的人有时也会打这张牌。最伟大的智慧，有时候看上去也不见得很聪明。你不必真的很蠢，只不过是装傻而已。对蠢人明智，和对智者愚蠢一样，都没什么作用，因此要见人说人话，见鬼说鬼话。装蠢者并非蠢人，真蠢者才是。精致巧妙的愚蠢，比起简单笨拙的装傻，更是真正的愚蠢，因为只有智者才能达到这样的高度。要想倍受喜爱，你必须披上愚蠢动物的毛皮。

MAKE USE OF FOLLY.

The wisest person plays this card at times. Sometimes the greatest wisdom lies in seeming not to be wise. You need not be unwise, but merely affect unwisdom. To be wise with fools and foolish with the wise is of little use; speak to each in his own language. He is no fool who affects folly, but he is who suffers from it. Ingenious folly, rather than simple affect, is the true foolishness, since cleverness is at such a high pitch. To be well liked one must dress in the skin of the simplest of animals.

www.buhua.com
2004

容忍别人的嘲笑，但自己不要践行

PUT UP WITH MOCKERY BUT DO NOT PRACTICE IT YOURSELF.

容忍嘲弄，是一种谦恭的姿态；嘲弄他人，则会让你陷入困境。对别人的打趣逗乐大发雷霆，看上去就令人生厌。无拘无束的嘲弄令人愉快，容忍它方能显出你的雅量。显露自己的恼怒也会引起他人的不快。最好的办法是听之任之——这是避免戴上小丑帽的最为可靠的办法。最严肃的事曾经由于玩笑而引起。没有什么事比玩笑需要更多的机智和警觉。在你开玩笑之前，要弄清楚你所嘲笑的对象在多大程度上能够容忍它。

The first is a form of courtesy, the second may lead to embarrassment. To snarl at playful jokes seems beastly. Audacious mocking is delightful and to stand for it proves your power. To show oneself annoyed causes others to be annoyed. Best leave it alone — that is the surest way of avoiding fitting the fool's cap. The most serious matters have arisen out of jests. Nothing requires more tact and attention. Before you begin to joke know how far the subject of your joke is able to bear it.

www.buhua.com
2004

善始善终

有些人做事总是虎头蛇尾。他们总在设计，却从不实施。有一些模棱两可的人——他们不可能因为他们的有始无终而赢得美名。对他们来说，每件事情在第一步就结束了。在有些人的身上，这种毛病起因于急躁，这正是西班牙人的缺点，正如耐性是比利时人的美德一样。后者善始善终，而前者则总是草草收场。他们挥汗如雨，直到障碍被克服，但他们只满足于此。他们不懂得如何一鼓作气，赢得最后的胜利。他们证明了自己能做，只是不愿意做。这表明，他们要么是没能力，要么是不可靠。如果你着手去做的是桩好事，为什么不做完呢？如果是桩坏事，那为什么又要着手呢？如果你明智的话，那么就打死你的猎物吧——而不要仅仅满足于把它从隐藏的地方赶出来。

PUSH ADVANTAGES.

Some put all their strength in the commencement and never carry a thing to conclusion. They invent but never execute. These be ambiguous spirits — they obtain no fame for they sustain no game to the end. Everything ends at the first stop. In some that arises from impatience, which is the failing of the Spaniards, as patience is the virtue of the Belgians. The latter bring things to an end, the former come to an end with things. They sweat away until the obstacle is overcome, but then they are content — they do not know how to push the victory home. They prove that they can but will not. This shows that they are either incapable or unreliable. If the undertaking is good, why not finish it? It is bad, why undertake it? Strike down your quarry, if you are wise — do not be content merely to flush it out.

我回来啦！

不要总是做鸽子

要让毒蛇的狡猾和鸽子的率真交替为用。欺骗一个诚实率直的人，最容易不过。从不说谎者不难相信别人，不会骗人者容易被人所骗。被人欺骗并非总是由于愚蠢，也可能是因为十足的善良。有两类人能够保护自己免受伤害：一种人是通过经历而得到相关的教训，代价当然是自己付出的；另一种人通过观察而洞悉了欺骗的奥秘，代价则是由别人支付。使用怀疑时的审慎和使用圈套时的精明要旗鼓相当，绝对不必心地太好，以至于使得别人能够对你使坏。把鸽子和毒蛇结合到你的身上，这并不像个妖怪，而更像个奇才。

DO NOT BE TOO MUCH OF A DOVE.

Alternate the cunning of the serpent with the candor of the dove. Nothing is easier than to deceive an honest man. He believes in much who lies about nothing; he who does no deception has much confidence. To be deceived is not always due to stupidity, it may arise from sheer goodness. There are two sets of people who can guard themselves from injury: those who have learned by experiencing it at their own cost and those who have observed it at the cost of others. Prudence should use as much suspicion as subtlety uses snares, and none need be so good as to enable others to do him ill. Combine in yourself the dove and the serpent, not as a monster but as a prodigy.

www.buhua.com
2004

制造一种负债感

有些人善于把受惠转变成施惠，当他们接受恩惠的时候，看上去似乎是在施与恩惠——或者让人们认为是这样。有些人精明得很，能通过求人而得到尊敬，用别人的嘉许来换取自己的好处。他们处理事情的手法很巧妙，以至于当他们接受别人效劳的时候看上去似乎是在为别人效劳。他们以非凡的技巧把债务关系颠倒了过来，或者至少是让谁是债主变得很可疑。他们通过赞美来换取最好的东西，把他们所表达的愉悦变成一种讨人喜欢的荣耀。他们通过自己的谦恭让人受宠若惊，使得人们为了他们自己本应该感激的事情而对他们感恩戴德。就这样，主动语态的“感激”变成了被动语态，由此证明，他们更多的是政治家，而不是语法家。这是一招巧妙的手腕，但更高的手段是：识破它，用他们自己口袋里的钱来偿还这笔人情债，并让这笔钱重新落到你自己的手里。

CREATE A FEELING OF OBLIGATION.

Some transform favors received into favors bestowed, and seem — or let it be thought — that they are doing a favor when receiving one. There are some so astute that they get honor by asking, and buy their own advantage with applause from others. They manage matters so cleverly that they seem to be doing others a service when receiving one from them. They transpose the order of obligation with extraordinary skill, or at least render it doubtful who has obliged whom. They buy the best by praising it, and make a flattering honor out of the pleasure they express. They oblige by their courtesy, and thus make people beholden for what they themselves should be indebted. In this way the conjugate 'to oblige' in the active instead of in passive voice, thereby proving themselves better politicians than grammarians. This is a subtle piece of finesse, but even greater is to perceive it, and to retaliate on such fools' bargains by paying in their own coin, and so come into your own again.

这家伙！提供
多少就业机会！

要有独创而别具一格的观点

HAVE ORIGINAL AND OUT-OF-THE-WAY VIEWS.

这是才能出众的标志。我们总是很少想起某个从不反驳我们的人；这并不表明他爱我们，而是表明他爱自己。不要被阿谀奉承所欺骗，并为此付出代价，而是要谴责它。此外，你可能因为被某些人批评而受人信任，尤其是那些好人说他们很坏的人。相反，如果我们所做的事情能取悦每个人的话，那倒是要感到不安，因为这是一个信号，表明这些事情毫无价值。完美总是为少数人而存在的。

These are signs of superior ability. We do not think much of someone who never contradicts us; that is not a sign he loves us but rather that he loves himself. Do not be deceived by flattery and thereby have to pay for it, rather condemn it. Besides, you may be given credit for being criticized by some, especially if they are those of whom the good speak ill. On the contrary, it should disturb us if our affairs please everyone, for that is a sign that they are of little worth. Perfection is for the few.

不要满足人们并未提出的要求

如果他们并没有对此提出要求，你给出的比需要的更多就是一种愚蠢行为。在有必要辩解之前就为自己辩解，其实就是在控告自己。在完全健康的时候放血，就等于暗示自己有病。一次意料之外的申辩，会唤醒正在沉睡的怀疑。精明之人即使意识到别人的怀疑，也不会有任何表露，否则就等于不打自招。通过自己行为的正直来消除人们的怀疑，方是上策。

NEVER OFFER SATISFACTION UNLESS IT IS DEMANDED.

And if they do demand it, it is a kind of crime to give more than necessary. To excuse oneself before there is occasion is to accuse oneself. To draw blood in full health gives the hint to ill will. An excuse unexpected arouses suspicion from its slumbers. Nor need a shrewd person show himself aware of another's suspicion, which is equivalent to seeking out offense. He had best disarm distrust by the integrity of his conduct.

www.buhua.com
2004
z z z

懂得多一点，活得少一点

有些人说的则相反。安逸悠闲好于事务缠身。除了时间，没有什么东西真正属于我们，就算你一无所有，时间总还是有的。把宝贵的生命浪费在机械刻板的工作上，或者浪费在大量太重要的工作上，同样都是不幸。不要让事情堆积如山，并因此而羡慕别人的悠闲，否则你会让生活变得复杂，并让自己筋疲力尽。有些人希望把同样的原则应用在知识上，但一个人如果没有知识的话，他就没有真正的生活。

KNOW A LITTLE MORE, LIVE A LITTLE LESS.

Some say the opposite. To be at ease is better than to be at business. Nothing really belongs to us but time, which you have even if you have nothing else. It is equally unfortunate to waste your precious life in mechanical tasks or in a profusion of too important work. Do not heap up occupation and thereby envy, otherwise you complicate life and exhaust your mind. Some wish to apply the same principle to knowledge, but unless one knows one does not truly live.

不要附和最后说话的人

有的人遵从他们最后听到的话，并因此而走向失去理性的极端。他们的感情与愿望都是蜡做的，最后来的人会在上面留下自己的封印，并擦除先前的所有印痕。这些人绝不会得到任何东西，因为每样东西他们都很快就失去了。每个人都会给他们染上自己的色彩。不能把他们视为心腹知己，他们整个一生都是长不大的孩子。由于感情和意志的这种变化无常，他们总是踉踉跄跄，是意志和思想上的跛子，从路的一侧蹒跚着走向另一侧。

DO NOT GO WITH THE LATE SPEAKER.

There are people who go by the latest thing they have heard and thereby go to irrational extremes. Their feelings and desires are made of wax; the last comer stamps them with his seal and obliterates all previous impressions. These people never gain anything, for they lose everything so soon. Everyone dyes them with his own color. They are of no use as confidants; they remain children their whole life. Owing to this instability of feeling and volition, they stumble along, crippled in will and thought, tottering from one side of the road to the other.

不要从本该在生命结束的时候干的事情上开始生活

很多人一开始就追求享乐，而把忧虑推到最后；但本质的东西应该首先出现，然后才是附属的东西，如果还有余地的话。有些人希望在战斗之前就取得胜利。还有人从学习无足轻重的事情开始，而丢下那些能给他们带来名声、让他们实现生活目标的学问。另一些人则在他们正要获得财富的时候突然从现场消失了。无论对于知识还是对于生活，方法才是本质的东西。

NEVER BEGIN LIFE WITH WHAT SHOULD END IT.

Many take amusement at the beginning, putting off anxiety to the end; but the essential should come first and accessories afterwards if there is room. Others wish to triumph before they have fought. Others again begin with learning things of little consequence and leave studies that would bring them fame and gain to the end of life. Another is just about to make his fortune when he disappears from the scene. Method is essential for knowledge and for life.

何时该反着听话

当他们语带恶意的时候，就该反着听话。对某些事情他们总是反着说：他们说的“不”其实就是“是”，他们说的“是”其实是“不”。如果他们说某个东西很坏，那其实是最高的赞美。正因为他们想得到这个东西，所以他们才对别人贬低它。要赞美一件东西，并不总是说它好。因为有些人通过说某个东西坏，从而避免说它好。对那些认为任何东西都不坏的人来说，其实就是任何东西都不好。

WHEN TO TURN CONVERSATION AROUND.

When they talk malice. With some everything goes in reverse: their no is yes and their yes is no. If they speak ill of something it is the highest praise. For what they want for themselves they depreciate to others. To praise a thing is not always to speak well of it. For some avoid praising what's good by praising what's bad. Nothing is good for him for whom nothing is bad.

臭狗屎！

使用人力仿佛神力不在，利用神力宛如人力乌有

USE HUMAN MEANS AS IF THERE WERE NO DIVINE ONES, AND DIVINE MEANS AS IF THERE WERE NO HUMAN ONES.

这是一条铁律，无需置评。

A masterful rule, which needs no comment.

管
击
手
腕
人
抓

既不要完全属于自己，也不要完全属于别人

二者都是暴政的粗鄙形态。想要全部属于自己，就是想让自己拥有全部。这样的人不会做出最小的让步，也不会让自己的安逸失去分毫。他们很少感激什么，只是依靠自己的幸运，而他们所依靠的通常并不牢固。有时候属于他人是合算的，那样别人或许也属于我们。担任公职的人，只不过是公众的奴仆，既不必比这更多，也不会比这更少。正如那位老妇人对哈德良[①]所说的：让一个人放弃职位，就是让他放下担子。另一方面，有些人则完全为他人而活着——这是愚蠢（愚蠢总是飞过了头），而且是一种最不幸的生活方式。没有一日、没有一时是他自己的。他们是如此彻头彻尾地属于他人，以至于人们或许会把他们称作所有人的奴隶。这甚至可以用之于知识领域，有些人对别人的事情无所不知，而对关乎到自己的事情却一无所知。你如果明智就应该知道：别人之所以找你，不过是因为你是他们的利益之所在、之所由，否则，他们不会找你。

① 哈德良(76-138)，罗马皇帝，罗马“五贤君”之一。117-138年在位。

NEITHER BELONG ENTIRELY TO YOURSELF NOR ENTIRELY TO OTHERS.

Both are mean forms of tyranny. To desire to be all for oneself is the same as desiring to have all for oneself. Such people will not yield the least bit or lose the smallest portion of their comfort. They are rarely beholden, lean on their own luck, and their crutch generally breaks. It is convenient at times to belong to others so that others may belong to us. And he that holds public office is no more nor less than a public slave; let a man give up both berth and burden, as the old woman said to Hadrian. On the other hand, some people are all for others — this is folly, which always flies to extremes, and in this case in a most unfortunate manner. No day, no hour, is their own. They so much belong to others that they may be called slaves to all. This applies even to knowledge, where a person may know everything for others and nothing for himself. A shrewd person knows that others, when they seek him, do not seek him but their advantage in him and by him.

www.buhua.com
2004

别解释得太清楚

大多数人并不尊重他们已经懂得的东西，而对自己不甚了然的东西敬佩有加。要让某种东西被人珍视，就要让它代价昂贵；不被人理解的东西，人们对它的估价更高。要想让自己得到很高的评价，你就要表现得比与你打交道的人所期望的更明智、更审慎。然而在这方面也要适可而止，毋过犹不及。虽说跟明智之士打交道只需坚守常识就行，但对大多数人来说，略费苦心也是必要的。别给他们批评的机会——让他们忙于揣摩你的意思而无暇他顾。许多人赞美一件东西，却说不出为何赞美——如果有人问及的话。原因就在于，他们把未知的东西奉若神明，之所以赞美它是因为听到它受到别人的赞美。

DO NOT EXPLAIN TOO MUCH.

Most people do not esteem what they understand and venerate what they do not see. To be valued things should cost dear; what is not understood becomes overrated. You have to appear wiser and more prudent than is required by the people you are dealing with if you want to give a high opinion of yourself. Yet in this there should be moderation and no excess. And though with sensible people common sense holds its own, with most people a little elaboration is necessary. Give them no time for criticizing — occupy them with discerning your meaning. Many praise a thing without being able to tell why, if asked. The reason is that they venerate the unknown as a mystery, and praise it because they hear it praised.

不要轻视不幸，无论它多么渺小

祸不单行，它们总是成群结队，就像一连串的好运。无论福祸，通常都会去找自己的同伙。因此，趋利避害，人人皆然。就连天真无邪的鸽子也会成群结队地飞向最白的围墙。不幸之人，万事皆糟——他自己，他的言辞，他的运气。当不幸沉睡的时候，不要唤醒它。一次闪失不过是小事一桩，但一些致命的损失或许会接踵而至，直至你不知何处是尽头。正如没有幸福是完美的，糟糕的运气也并非全然不幸。以耐心面对来自上天的灾祸，以审慎应对来自尘世的不幸。

NEVER DESPISE AN EVIL, HOWEVER SMALL.

They never come alone, they are linked together like pieces of good fortune. Fortune and misfortune generally go to find their fellows. Hence all avoid the unlucky and associate with the fortunate. Even the doves with all their innocence resort to the whitest walls. Everything fails with the unfortunate — himself, his words, and his luck. Do not wake misfortune when she sleeps. One slip is a little thing, yet some fatal loss may follow it till you do not know where it will end. For just as no happiness is perfect, so no piece of bad luck is complete. Use patience with what comes from above, prudence with that from below.

www.buhua.com
2004

行善一次勿需太多，但要经常

你不能付出太多，以至于超出了回报的可能。付出太多的就不是付出，而是贩卖。也不要把别人的感激消耗殆尽，只剩下残渣，受惠者眼见得全部回报几无可能，便会中断往来。对许多人，给予他们超出其承载能力的恩惠是不必要的，那样就会完全失去他们。既然无法报答，他们也就会知难而退，宁愿成为你的敌人，也不愿成为永久的债务人。雕像决不希望见到塑造自己的雕刻家，受惠之人也不愿意看到自己的恩人老在面前晃来晃去。此处大有奥妙：给予某些价格甚微但别人又迫切想得到的东西，能够得到更多的尊敬。

DO GOOD A LITTLE AT A TIME, BUT OFTEN.

One should never give beyond the possibility of return. He who gives much does not give but sells. Nor drain gratitude to the dregs, for what the recipient sees all return is impossible he breaks off correspondence. With many people it is not necessary to do more than overburden them with favors to lose them altogether; they cannot repay you, and so they retire, preferring rather to be enemies than perpetual debtors. The idol never wishes to see before him the sculptor who shaped him, nor does the benefited wish to see his benefactor always before his eyes. There is a great subtlety in giving what costs little yet is much desired, so that it is esteemed the more.

有备而往

有备而往可以战胜粗鲁无礼、背信弃义、专横独断，以及其他种种蠢行。世界上这样的蠢行很多，审慎就在于避免与之不期而遇。每天要在注意之镜的前面用防卫的武器装备自己。这样你就可以打垮蠢行的进攻。要时刻做好准备，不要让你的名誉暴露在粗俗的意外事件的攻击之下。用审慎武装自己，你就不可能被鲁莽缴械。人类交往的大道举步维艰，因为那里充满了让我们的名誉颠簸摇晃的沟沟坎坎。最好是抄小路，把尤利西斯作为精明机智的楷模。在此类事情上，假装误会是极有价值的做法。辅之以文雅有礼，它对我们的帮助就无处不在，而且常常是摆脱困境不二法门。

GO PREPARED.

Go armed against discourtesy, faithlessness, presumption, and all other kinds of folly. There is much of it in the world, and prudence lies in avoiding meeting with it. Arm yourself each day before the mirror of attention with the weapons of defense. Thus you will beat down the attacks of folly. Be prepared for the occasion, and do not expose your reputation to vulgar contingencies. Armed with prudence, a person cannot be disarmed by impertinence. The road of human intercourse is difficult, for it is full of ruts that may jolt our reputation. Best to take the byway, taking Ulysses as a model of shrewdness. Feigned misunderstanding is of great value in such matters. Aided by politeness it helps us over all, and is often the only way out of difficulties.

防
装傻
察言
永不輕信
小心
保持距离

不要让事情达到断裂点

因为结果总是给我们的名誉带来损害。每个人都有其价值，如果不是作为朋友，便是作为敌人。能对我们好的人并不多，而几乎任何人都能给我们带来伤害。从朱庇特跟甲虫反目的那天起，就连他的神鹰都再也没睡过一个安稳觉，哪怕是在朱庇特的怀里。隐藏的仇敌，利用公开敌人的手爪来煽风点火，而他们则埋下伏兵，伺机而动。被激怒的朋友，到头来成了最怀恨的敌人。他们用别人的过错来掩盖自己的弱点。每个人都按照在自己看来的样子去谈论事情，而事情似乎也像他们希望呈现出的样子。每个人都责备我们一开始就缺乏远见，到头来又缺乏审慎，自始至终都轻率鲁莽。然而，如果破裂确实不可避免的话，那么最好让人们把它当作一次对友谊的怠慢而予以原谅，而不要通过一次情绪爆发或怒火中烧来实现决裂。这正是对那条关于如何明智退出的格言的良好应用。

NEVER LET MATTERS COME TO A BREAKING POINT.

For our reputation always comes out injured. Everyone may be of importance as an enemy if not as a friend. Few can do us good, almost any can do us harm. In Jove's bosom itself even his eagle never nestles securely from the day he has quarreled with a beetle. Hidden foes use the paw of the declared enemy to stir up the fire, and meanwhile they lie in ambush for such an occasion. Friends provoked become the bitterest of enemies. They cover their own failings with the faults of others. Everyone speaks as things seem to him, and things seem as he wishes them to appear. Everyone will blame us at the beginning for want of foresight, at the end for lack of patience, at all times for imprudence. If, however, a breach is inevitable, let it be rather excused as a slackening of friendship than by an outburst or wrath. This is good application of the saying about a good retreat.

找人替你分忧

这样，你才不会孤苦伶仃。即使身处险境，也不要把所有憎恨的重担都自己扛。有些人自认为身处高位，就可以赢得成功的全部荣誉，到头来却发现自己不得不背上失败的全部耻辱。这样一来，既没人原谅他们，也没人分担责任。无论是厄运还是暴民，都不敢贸然和两个人作对。因此，高明的大夫在自己无力回天的时候，就会以会诊的名义物色某位同行，来帮着他抬尸首。必须有人分担重负和悲伤，因为独自承担的话，不幸就会以双倍的力量落在你的肩上。

FIND SOMEONE TO SHARE YOUR TROUBLES WITH.

You will never be all alone, even in dangers, nor bear all the burdens of hate. Some think by their high position that they can carry off the whole glory of success, and find that they have to bear the whole humiliation of defeat. In this way they have no one to excuse them, no one to share the blame. Neither fate nor the mob are so bold against two. Hence the wise physician, if he has failed to cure, looks out for someone who, under the name of a consultation, may help him carry out the corpse. Share weight and woe, for misfortune falls with double force on him that stands alone.

www.buhua.com
2004

预见伤害，化敌为友

对待侮辱，避免比报复更明智。化生死对头为心腹知己，或者将伺机攻击我们的人转变为我们的荣誉卫士，是一种非凡的智巧。这对于搞懂怎样让别人心怀感激大有帮助，因为时刻充满感激的人，就没有时间去伤害别人。转忧为乐，方为大智。努力让敌意成为你的密友。

ANTICIPATE INJURIES AND TURN THEM INTO FAVORS.

It is wiser to avoid than to revenge them. It is an uncommon piece of shrewdness to change a rival into a confidant, or transform into guards of honor those who were aiming to attack us. It helps much to know how to oblige, for he leaves no time for injuries who fills time up with gratitude. It is true savoir faire, to turn anxieties into pleasures. Try and make a confidential relation out of ill will itself.

www.buhua.com
2004

我们并不完全属于别人，别人也不完全属于我

亲朋好友乃至最亲密的关系，都不足以达到不分彼此的程度。充分信任完全不同于尊敬。最亲密的人之间亦有其例外，不如此，友谊的法则就会遭到破坏。朋友总会保留属于自己的那份秘密，即便是儿子，也会对父亲有所隐瞒。某些事情，我们对此人闭口不谈，而对彼人却开诚布公，反之亦然。这样说来，一个人袒露一切，同时又隐瞒一切，只不过所针对的人有所不同而已。

WE BELONG TO NO ONE AND NO ONE TO US, ENTIRELY.

Neither relationship nor friendship nor the most intimate connection is sufficient to effect this. To give one's whole confidence is quite different from giving one's regard. The closest intimacy has its exceptions, without which the laws of friendship would be broken. The friend always keeps one secret to himself, and even the son always hides something from his father. Some things are kept from one that are revealed to another and vice versa. In this way one reveals all and conceals all, by making a distinction among the persons with who we are connected.

www.buhua.com
2004

不要把蠢行进行到底

许多人从失误中创造出了义务，因为他们走上了错误的小路，于是认为继续走下去便是对品格力量的证明。在内心里，他们为自己的错误而扼腕叹憾，而在表面上，他们却为自己的错误开脱。一开始犯错的时候，人们认为他们是粗心大意，到最后，人们便认为他们愚不可及。不管是轻率的许诺，还是错误的决定，都没有真正的约束力。然而有些人总是继续他们的愚蠢，宁愿做一个坚定不移的傻瓜。

DO NOT FOLLOW UP A FOLLY.

Many make an obligation out of a blunder, and because they have entered the wrong path they think it proves their strength of character to go on in it. Within they regret their error, while outwardly they excuse it. At the beginning of their mistake they were regarded as inattentive, in the end as fools. Neither an unconsidered promise nor a mistaken resolution are really binding. Yet some continue in their folly and prefer to be constant fools.

放心吧孩子！倾家荡产也支持你追星！
对！周杰伦要是不娶你，爷爷就不活了！！
对！

学会遗忘

遗忘，与其说是靠技巧，还不如说是凭运气。我们记得最牢的，总是那些最应该忘掉的事情。记忆，不仅难以驾驭，在我们最需要的时候把我们留在困境中，而且还很愚蠢，总是把它的鼻子伸进那些我们不希望被探测的地方嗅来嗅去。在痛苦的事情上它很积极，但在我们回忆赏心乐事的时候它总是漫不经心。对烦恼来说，唯一的治疗方法常常是忘掉它，而我们所忘掉的一切，恰恰是这剂良药。然而，一个人还应该培养良好的记忆习惯，因为它既能把生活变成天堂，也能使之成为地狱。幸福只是一个例外，只有那些天真无邪地享受他们的简单快乐的人才会得到。

BE ABLE TO FORGET.

It is more a matter of luck than of skill. The things we remember best are those better forgotten. Memory is not only unruly, leaving us in thc lurch when most needed, but stupid as well, putting its nose into places where it is not wanted. In painful things it is active, but neglectful in recalling the pleasurable. Very often the only remedy for the trouble is to forget it, and all we forget is the remedy. Nevertheless one should cultivate good habits of memory, for it is capable of making existence a paradise or an inferno. The happy are an exception who enjoy innocently their simple happiness.

有滋有味的东西不必自己拥有

如果它们是别人的东西，你从中得到的享受胜过自己的。拥有者不过是第一天享受它们的好处，而在其余的所有时间里，它们都是为别人而准备的。从别人的财产中，你可以得到双倍的享受，既不用担心弄坏它，又能得到新奇的乐趣。每件有滋有味的东西，都因为不曾拥有而滋味更佳——即使是来自别人井里的水，味道也像甘露一样。占有，总是妨碍享受，徒增烦忧，无论你借出还是厮守。这些东西，你要么是为别人保管，要么是不让他人染指，除此之外你什么也得不到，由此，你得到的更多的是敌人，而不是朋友。

MANY THINGS OF TASTE ONE SHOULD NOT POSSESS ONESELF.

One enjoys them better if they are another's rather than one's own. The owner has the good of them first day, for all the rest of the time they are for others. You take a double enjoyment in other men's property, being without fear of spoiling it and with the pleasure of novelty. Everything tastes better for having been without it — even water from another's well tastes like nectar. Possession hinders enjoyment and increases annoyance, whether you lend or keep. You gain nothing except keeping things for or from others, and by this means gain more enemies than friends.

家
工
作

不要有粗心大意的时候

命运之神喜欢搞恶作剧，它总是把机会积累起来，出其不意地堆放在我们面前。我们的智慧、审慎、勇气，甚至还有我们的美貌，必须时刻准备接受这样的考验。因为对信任粗心大意的日子，也就是失去信任的日子。谨慎细心总是在你最需要它的时候不见踪影。正是粗心大意让我们跌倒在毁灭之中。因此，在毫无准备的时候把完美置于考验之下是一招军事策略。耀武扬威的日子被人注目，被允许通过，但这个日子要选在最意想不到的时候，这样就可以让勇气接受最严格的检验。

HAVE NO CARELESS DAYS.

Fate loves to play tricks, and will heap up chances to catch us unawares. Our intelligence, prudence, and courage, even our beauty, must always be ready for trial. For their day of careless trust will be that of their discredit. Care always fails just when it was most wanted. It is thoughtlessness that trips us up into destruction. Accordingly, it is a piece of military strategy to put perfections to their trial when unprepared. The days of parade are watched and are allowed to pass, but the day is chosen when least expected so as to put valor to the severest test.

给手下人设置困难的任务

许多人在不得不应对困难的时候立刻就证明了自己是能干的，正像对葬身鱼腹的恐惧把一个人造就成了泳者一样。就这样，许多人发现了自己的勇气、知识和机智，要不是有这种机会的话，这些品质将会永远埋藏在他们的畏缩被动之下。危险的环境，就是一个人独立创造自我的机会，如果一个精神高贵的人看到荣誉受到威胁，他会做千百人的工作。天主教徒伊莎贝拉女王就深知这一生活准则（还有所有别的准则），把这种精明的恩惠赐予给了那位为自己赢得美名的大船长[①]，以及许多其他名垂青史的人。凭借这种伟大的技艺，她造就了很多伟大的人物。

① 这里指的是西班牙将军弗朗西斯科·弗尔南德斯·德·科多巴(1453-1515)，他曾指挥西班牙军队抵抗法国国王查理八世。

SET DIFFICULT TASKS FOR THOSE UNDER YOU.

Many have proved themselves able at once when they had to deal with difficulty, just as fear of drowning makes a person into a swimmer. In this way, many have discovered their own courage, knowledge, or tact, which but for the opportunity would have been forever buried beneath their lack of initiative. Dangerous situations are the occasions to create a name for oneself, and if a noble mind sees honor at stake, he will do the work of thousands. Queen Isabella the Catholic knew well this rule of life (as well as all the others) and to a shrewd favor of this kind of Great Captain won his fame, and many others earned an undying name. By this great art she made great men.

命令你把这道
代数题帮我做出来！

过善则恶

不要因为过于善良(换句话说,也就是从不生气)而适得其反。没有脾气的人几乎不能被视为人。这并不总是由于怠惰,而是因为十足的无能。有时候脾气火爆可以显示个性,就连小鸟也会嘲弄稻草人。有苦有甜,方是至味。一味的甘甜,那是给孩子和傻瓜准备的食物。由于太过善良而陷入这样的麻木冷漠,实为大恶。

DO NOT BECOME BAD FROM SHEER GOODNESS.

That is, by never getting angry. Such people without feeling are scarcely to be considered human. It does not always arise from laziness, but from sheer inability. To feel strongly on occasion shows personality; birds soon mock at the scarecrow. It is a sign of good taste to combine bitter and sweet. All sweets is diet for children and fools. It is a great evil to sink into such insensibility out of too great goodness.

www.buhua.com
2004

温言软语，态度平和

利箭刺穿身体，凌辱伤透心灵。甜点心让人齿颊生香。懂得如何以空言换取实惠，是生活中的大智大巧。大多数东西是用甜言美语换来的，凭此，你可以排除不可能之事。因此，我们拿空言作交易，高贵的谈吐能产生勇气和力量。要始终注意你的言辞，这样就连你的敌人也会欣然接受。要取悦于人，须态度平和。

SILKEN WORDS, SUGARED MANNERS.

Arrows pierce the body, insults the soul. Sweet pastry perfumes the breath. It is a great art in life to know how to sell wind. Most things are paid for in words, and by them you can remove impossibilities. Thus we deal in air, and a royal breath can produce courage and power. Always have your words, so that even your enemies enjoy them. To please one must be peaceful.

www.buhua.com
2004

蠢人后做的事，智者先做

二者做的是同样的事——唯一的不同就在于他们做事的时机：一者是在正确的时间，而另一者是在错误的时间。颠三倒四、头脑混乱地着手工作的人，会这样继续下去，直到最后。本该挠头，他却抓脚，颠上倒下，变左为右，在所有方面他的动作都是不成熟的。只有一种办法可以让他转到正确的方向来，那就是强迫他去做他迟早要做的事，这样他就会心甘情愿地去做，并因此赢得尊敬。

THE WISE DO AT ONCE WHAT THE FOOL DOES LATER.

Both do the same thing — the only difference lies in the time they do it: the one at the right time, the other at the wrong. Who starts out with his mind topsy-turvy will so continue till the end. He catches by the foot what he ought to knock on the head, he turns right into left, and in all his acts is immature. There is only one way to turn him in the right direction, and that is to force him to do what he might have done sooner or later, so he does it willingly and gains honor thereby.

你今天必须出去找工作了!!

利用你的新位置

当你是位新人的时候，就会受到重视。新鲜事物，人人喜欢，因为它难得一见，令人耳目一新。一位资质平平的新人比司空见惯的杰出人士更受人关注。天纵英才会因为不断使用而日渐消磨，变得老气横秋。然而，你应该知道：新奇光环，其命亦短。四天之后，关注不再。因此，要懂得利用人人激赏的初熟之果，在转瞬即逝的喝彩欢呼声中，抓住一切可以利用的机会。一旦新奇的狂热结束，激情就会冷却，人们对新鲜事物的欣赏就会转换为对司空见惯的嫌恶。请相信：万物皆有其时，风光转瞬即逝。

MAKE USE OF THE NOVELTY OF YOUR POSITION.

For people are valued while they are new. Novelty pleases all because it is uncommon, taste is refreshed, and a brand new mediocrity is thought more of than accustomed excellence. Ability wears away by use and becomes old. However, know that the glory of novelty is short lived. After four days respect is gone. Accordingly, learn to utilize the first fruits of appreciation, and seize during the rapid passage of applause all that can be put to use. For once the heat of novelty is over, the passion cools and the appreciation of novelty is exchanged for distaste at the customary. Believe that everything has its season, which soon passes.

www.buhua.com
2004

众人所好之事，不要一人反对

一件东西，既然被这么多人喜爱，必定有其可取之处——即便无从解释，但它的确深受欢迎。怪僻总是招致憎恨，如果它是错的，还会遭人嘲讽。你简直是在摧毁人们对你的品味的尊敬，这更甚于对你所谴责的对象的损害，而且，人们不再理会你，还有你糟糕的品味。如果你不能在一件事情中找到优点，你最好还是藏拙，而不要滥加指责。依照通行的法则，恶劣的品味源自知识的匮乏。众口一词的东西，要么就是这样，要么人们愿意这样。

DO NOT CONDEMN ALONE THAT WHICH PLEASES ALL.

There must be something good in a thing that pleases so many — even if it cannot be explained it is certainly enjoyed. Peculiarity is always hated and, when in the wrong, laughed at. You simply destroy respect for your taste rather than do harm to the object of your blame, and are left alone, you and your bad taste. If you cannot find the good in a thing, hide your incapacity and do not damn it right away. As a general rule bad taste springs from want of knowledge. What all say, is so, or will be so.

www.buhua.com
2004

不做无把握之事

IN EVERY OCCUPATION, IF YOU KNOW LITTLE STICK TO THE SAFE PATH.

这样的话，人们即使不认为你精明，但会认为你可靠。另一方面，一个受过良好训练的人，当然可以投身自己的爱好，并且游刃有余。如果你一无所知还要去冒险，那就等于自取灭亡。跟着习惯走，萧规曹随，往者可鉴。那些知识浅薄的人，应该遵循这条康庄大道。然而不管怎样，有知也好，无知也罢，安全稳妥总比别出心裁更为明智。

If you are not respected as subtle, you will be regarded as sure. On the other hand, someone well trained can plunge in and act as he pleases. To know little and yet seek danger is no different than to seek ruin. Follow the right hand, for what has gone before can be followed after. Let those with little knowledge keep to the king's highway, and in every case, knowing or unknowing, security is shrewder than uniqueness.

www.buhua.com
2004

以谦恭为价格卖东西

这样你就让人感激不尽。一位感兴趣的买主，其出价跟心存感激的恩惠接受者的回礼不可同日而语。谦恭并不送出真正的礼品，却把人置于义务的约束之下，慷慨是一大笔人情债。对正直之士来说，没有比白送给他的东西价格更昂贵的了。你实际上把这件东西以双倍的价格卖给他两次：一次是为它的实际价值付账，另一次是为你的礼貌买单。同时，有一点也是真的，对粗俗之人来说，慷慨就是胡言乱语，因为他们听不懂良好教养的语言。

SELL THINGS WITH A TARIFF OF COURTESY.

You oblige people most that way. The bid of an interested buyer will never equal the return gift of a grateful recipient of a favor. Courtesy does not really make presents, but lays people under obligation, and generosity is the great obligation. To the right-minded nothing costs more dear than what is given to him. You sell it to him twice and for two prices: one for the value, one for the politeness. At the same time, it is true that with vulgar souls generosity is gibberish, for they do not understand the language of good breeding.

笑得多喜兴，给拿一万！

了解你所交往的人的性格

COMPREHEND THE DISPOSITION OF THE PEOPLE YOU DEAL WITH.

这样你就会了解他们的意图。知其因必知其果;先知性格,后知动机。忧郁的人总是预见灾祸,搬弄是非者总是预见丑闻——没有善良的性格,邪恶就会出现在他们身上。一个被激情所左右的人,他所谈论的事情总是跟其本来面目大相径庭;说话的是他的激情,而不是他的理性。因此,他说的每一句话都是他的感觉或心情给他提的词,全都远离事实。要学会如何破译人们的表情,从外貌中洞悉灵魂。总是笑的人可以归类为傻瓜,从不笑的人则是虚伪的人。要留心闲言碎语——他要么是个胡言乱语者,要么就是密探。别指望从畸形人那里得到什么善意:他们通常要报复大自然,对大自然没什么敬意,因为大自然对他也是如此。美貌与愚蠢通常总是携手并肩,形影不离。

Then you will know their intentions. Cause known, effect known; beforehand in the disposition and after in the motive. The melancholy person always foresees misfortunes, the backbiter scandals — having no conception of the good, evil offers itself to them. A person moved by passion always speaks of things as different from what they are; it is his passion that speaks, not his reason. Thus each speaks as his feeling or his humor prompts him, and all far from the truth. Learn how to decipher faces and spell out the soul in the features. If someone always laughs set him down as foolish, if never as false. Beware of the gossip — he is either a babbler or a spy. Expect little good from the misshapen: they generally take revenge on nature, and do little honor to her, as she has done little to them. Beauty and folly generally go hand in hand.

嘿嘿！没有表情！猜猜我什么性格？
爱扯皮！

保持迷人的风度

这是精微礼节的魅力。利用你优雅品质的吸引力，比利用你的善行，更能吸引人们的善意。如果没有优雅的支撑，仅有优点是不够的。优雅，才是获得普遍接受的基本要素，也是支配他人最实用的手段。成为时髦人物是一件幸运的事，而技巧能襄助此事，因为技巧能扎根于大自然所偏爱的土壤之中。这样一来，善意就会生长发展，最终赢得普遍的喜爱。

BE ATTRACTIVE.

It is the magic of subtle courtesy. Use the magnet of your pleasant qualities more to attract goodwill than good deeds, but apply it to all. Merit is not enough unless supported by grace, which is the sole thing that gives general acceptance, and the most practical means of rule over others. To be vogue is a matter of luck, yet it can be encouraged by skill, for art can best take root on a soil favored by nature. There goodwill grows and develops into universal favor.

www.buhua.com
2004

只要不失庄重，可以参与游戏

不要总是装腔作势而成为一个令人讨厌的家伙——要想有高贵的风度，这是一句至理名言。为了赢得普遍的善意，你或许要在尊严方面稍作让步。你可以偶尔随波逐流，但不要逾越礼仪的界限。在公共场合把自己弄得像个傻瓜样的人，在私人生活里，人们也不会认为他是个谨言慎行之辈。你在一个快乐开心的日子所失去的东西，也许比在整个辛苦努力的一年当中所得到的还要多。尽管如此，你也不必总是离群索居，特立独行就是跟众人作对。少装假正经——把这种行为留给女人比较合适——即使是虔诚的一本正经也荒谬可笑。成为男子汉的最好办法，就是像个男子汉。一个女人装出一副男子气派，或许还显得像是卓尔不群，反之却并不然。

JOIN IN THE GAME AS FAR AS DECENCY PERMITS.

Do not always pose and be a bore — this is a maxim for gallant bearing. You may yield a touch of dignity to gain the general goodwill. You may now and then go where most go, yet not beyond the bound of decorum. He who makes a fool of himself in public will not be regarded as discreet in private life. One may lose more on a day of pleasure than has been gained during a whole year of labor. Still you must not always keep away; to be eccentric is to condemn all others. Still less act prudish — leave that to its appropriate sex — even religious prudery is ridiculous. Nothing so becomes a man as to be a man. A woman may affect a manly bearing as an excellence, but not vice versa.

www.buhua.com
2004

懂得如何更新你的性格，既由自然，亦以人工

人常说：性格脾气，七年一变。要使其在品味方面变得更好、更高贵。第一个七年之后，理性发蒙，随后的每一次改变，都要让一种新的优点为其增光添彩。注意这样的变化，助其发展，同时希望别的方面也能得到改良。因此，有许多人，当他们的地位和职业发生改变的时候，他们的言行举止也会随之改变。有时候，这种改变并不被人注意，直到它瓜熟蒂落、水到渠成。男人在二十岁的时候是只孔雀，三十岁是狮子，四十岁是骆驼，五十岁是毒蛇，六十岁是狗，七十岁是猿，到了八十岁，你就什么也不是。

KNOW HOW TO RENEW YOUR CHARACTER BOTH WITH NATURE AND WITH ART.

Every seven years the disposition changes, they say. Let it be a change for the better and for the nobler in your taste. After the first seven comes reason, with each succeeding luster let a new excellence be added. Observe this change so as to aid it, and hope also for betterment in others. Hence it happens that many change their behavior when they change their position or their occupation. At times the change is not noticed till it reaches the height of maturity. At twenty a man is a peacock, at thirty a lion, at forty a camel, at fifty a serpent, at sixty a dog, at seventy an ape, at eighty nothing at all.

www.buhua.com
2004

展示自己

这是对才干的彰显。每个人都有展示自己的合适时机——要善加利用，因为并非每天都能获得成功。有些才华横溢的人藏而不露，有些人则尽情展现。如果你打算展示的才能结合了多方面的天赋，那么你就会被视为奇迹。有的国家整个民族都善于展示自己；西班牙在这方面可谓登峰造极。创世之初，首先闪现的就是光。展示承载甚多，供应甚多，赋予事物以第二实体，尤其是当它结合了卓越品格的时候。上天给予世界以完美，同时也提供了展示完美的手段。即便是卓越的品格，也取决于环境，而且并非总是恰逢其时。不合时宜的炫耀卖弄，也就不合其地。最重要的品质，就是不要矫揉造作。否则的话，展示就是冒犯，因为这时候它就跟虚荣只有一步之遥，因此被人轻视。展示必须适度，以免粗俗，以免为智者所不齿。有时候，展示还在于沉默的力量，让卓越的品格在不经意间展现出来。聪明的掩饰，常常是最有效的夸耀，因为引而不发总是激起人们最大的好奇心。还有一种策略也是很巧妙的，那就是：不要一次把所有的优点展示殆尽，而是让人偷瞥一眼，随着时间的推移逐渐增多。每一次成就，应该是对更大成就的许诺；最初的鼓掌欢呼，只会让人们对结局的期待消失于无形。

DISPLAY YOURSELF.

It is the illumination of talents. For each there comes an appropriate moment — use it, for not every day comes to triumph. There are some dashing men who make a show with little and others who make a whole exhibition with much. If ability to display them is joined to versatile gifts, they are regarded as miraculous. There are whole nations given to display; the Spanish people take the highest rank in this. Light was the first thing to cause creation to shine forth. Display fills up much, supplies much, and gives a second existence to things, especially when combined with real excellence. Heaven, which grants perfection, also provides the means of display. Even excellence depends on circumstances and is not always opportune. Ostentation is out of place when it is out of time. More than any other quality it should be free of any affectation. If not, it is an offense, for it then borders on vanity and so on contempt. It must be moderate to avoid being vulgar, and any excess is despised by the wise. At times it consists of a sort of mute eloquence, a careless display of excellence. For a wise concealment is often the most effective boast, since the very withdrawal from view piques curiosity to the highest. It is a fine subtlety too, not to display one's excellences all at one time, but to grant stolen glances at it, more and more as time goes on. Each exploit should be the pledge of a greater, and applause at the first should only die away in expectation of its sequel.

过度表现
过
过
过
过
过
过
过
过

避免在所有事情上声名远扬

即便是优点，如果变得众所周知，也会成为缺点。怪癖所带来的名声总是受到谴责：特立独行之士总是让人敬而远之。即便是美貌，也会被过度的愚蠢所玷辱，它所吸引的每一次关注都会让人不快。不大光彩的怪癖则更是如此。然而总是有人想方设法让自己的恶劣行径为人所知，因为新奇的恶行可以获得狼藉的名声。即使是在知识学问上，缺乏适度也会流于空谈。

AVOID NOTORIETY IN ALL THINGS.

Even excellences become defects if they become notorious. Notoriety arises from eccentricity, which is always blamed: he that is singular is left severely alone. Even beauty is discredited by foolish excess, which offends by the very notice it attracts. Still more does this apply to discreditable eccentricities. Yet among the wicked there are some that seek to be known for seeking novelties in vice so as to attain to the fame of infamy. Even in matters of the intellect lack of moderation may degenerate into empty talk.

高尚

不要回应那些反驳你的人

你必须区别反驳究竟是来自狡诈之辈,还是来自粗俗之人。反驳并非总是出于固执,有可能出于狡诈。一定要注意这一点,因为在前一种情况下你没准会陷入困境,而在后一种情况下,则有可能陷入危险。对付密探的手段,莫过于小心。要防止人家撬开你的心灵之锁,最好的办法就是把小心的钥匙留在门锁的后边。

DO NOT RESPOND TO THOSE WHO CONTRADICT YOU.

You have to distinguish whether the contradiction comes from cunning or from vulgarity. It is not always obstinacy, but may be artfulness. Notice this, for in the first case one may get into difficulties, in the other into danger. Caution is never more needed than against spies. There is no countercheck to the picklock of the mind as to leave the key of caution in the inside lock of the door.

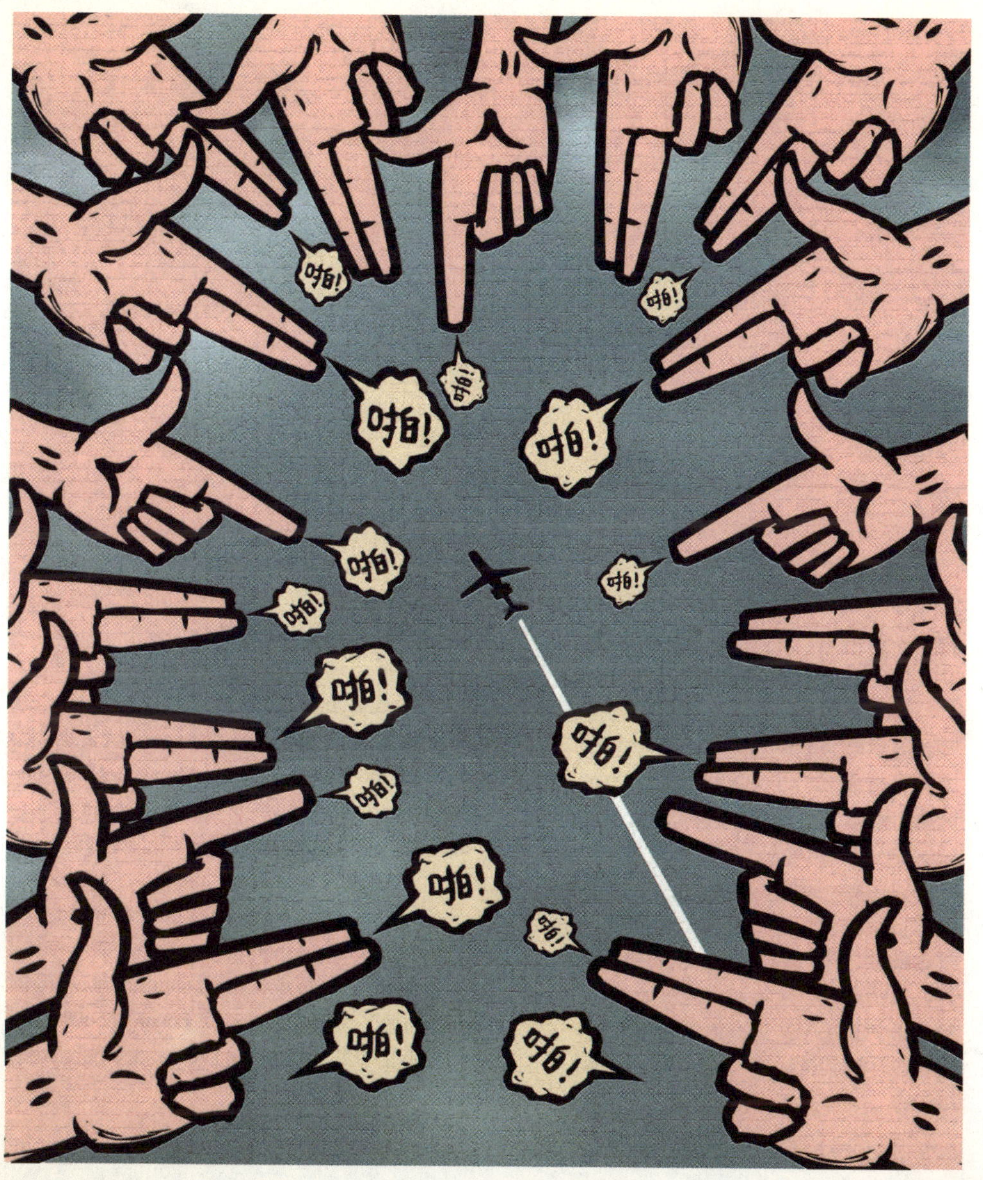
啪!
啪!
啪!
啪!
啪!
啪!
啪!
啪!
啪!
啪!
啪!
啪!
啪!
啪!
啪!

做一个值得信赖的人

值得尊敬的行为一旦结束，信任也就被收回，说话算数的人寥寥无几，效劳越多，回报越少——这就是当今世界的处事方式。有的国家，全民族都倾向于不诚实的行为；总让人害怕的，有些是背叛，有些是毁约，还有一些则是欺骗。然而对我们来说，这些劣行更应该是警示，而不是榜样。可怕的是，一看到这些卑劣的行为，会让我们自己的诚实消失得不见踪影。但一个值得尊敬的人决不能因为看到别人如何而忘记自己应该如何。

BE TRUSTWORTHY.

Honorable dealing is at an end, trusts are denied, few keep their word, the greater the service the poorer the reward — that is the way of the world nowadays. There are whole nations inclined to false dealing; with some treachery has always to be feared, with others breach of promise, with others deceit. Yet this bad behavior of others should be a warning to us rather than an example. The fear is that the sight of such unworthy behavior will override our integrity. But a person of honor should never forget what he is because he sees what others are.

跳！差不多
应该没事儿！

赢得智者的赏识

一位出类拔萃之士不冷不热的称许，比凡夫俗子的鼓掌欢呼更有价值——你不可能把谷壳冒的烟当作一顿饱餐。智者以理解力说话，他们的赞美能给人持久的满足。圣哲安提戈努斯把他赢得名声的戏台归到宙斯一个人，柏拉图称亚里士多德是自己的整个学校。有的人总是用子虚乌有的东西填饱自己的肚子，那怕它们只是乌合之众呼出的气息。就连君王也需要舞文弄墨之士，害怕他们的如椽之笔，甚于丑陋的女人害怕画家的笔。

FIND FAVOR WITH PEOPLE OF GOOD SENSE.

The tepid yes from a remarkable person is worth more than all the applause of the vulgar — you cannot make a meal of the smoke of chaff. The wise speak with understanding and their praise gives permanent satisfaction. The sage Antigonus reduced the theater of his fame to Zeus alone, and Plato called Aristotle his whole school. Some strive to fill their stomach albeit only with the breath of the mob. Even monarchs have need of authors, and fear their pens more than ugly women the painter's pencil.

好!
好!
好!
好!
好!
好!
好!

要有发现的才能

这是最高天才的明证，然而天才何时没有一点点疯狂呢？如果说发现是一种天赋之才的话，那么选择就是健全理性的标志。发现要靠特殊的恩典而偶然得到，而且非常罕见。许多人在一件事物被人发现的时候能追根问底，但最早发现它的，却是少数天才——在才能上出类拔萃，在时间上走在最前。新奇之物取悦众人，如果成功的话，会让拥有者得到双倍的信任。在有关判断力的事情上，新生事物是危险的，因为它们总是导致似非而是的结论；在天才所做的事情上，它们值得一切赞美。如果成功的话，二者同样都值得鼓掌喝彩。

HAVE THE GIFT OF DISCOVERY.

It is a proof of the highest genius, yet when was genius without a touch of madness? If discovery be a gift of genius, choice is a mark of sound sense. Discovery comes by special grace and very seldom. For many can follow up a thing when found, but to find it first is the gift of the few — the first in excellence and in age. Novelty flatters, and if successful gives the possessor double credit. In matters of judgement novelties are dangerous because they lead to paradox, in matters of genius they deserve all praise. Yet both equally deserve applause if successful.

少管闲事

不要多管闲事，那么你就不会被人小瞧。自重者，人重之。到场的时候，宁愿谨慎克制，而不要过分张扬。这样你就会成为人们所期待的人，并得到善待。不要不请自来，只有在有人召唤的时候才欣然前往。你主动承担的事，失败则招人责备，成功也无人感谢。多管闲事之人，最终总成笑柄，他们卷入时大言不惭，脱身时狼狈不堪。

DO NOT BE BURDENSOME.

Then you will not be slighted. Respect yourself if you would have others respect you. Be sooner sparing than lavish with your presence.You will thus become desired and so well received. Never come unasked and only go when sent for. If you undertake a thing of your own accord you get all the blame if it fails, none of the thanks if it succeeds. Those who do not mind their own business are always the butt of blame, and because they thrust themselves in without shame they are thrust out with it.

www.buhua.com
2004

不要因他人的厄运而自毁

留心那些深陷泥潭的人，注意他怎样寻求别人的帮助，为的是找到一个聊以自慰的同病相怜者。他要寻找某人来帮自己分担不幸，往往是那些在他顺境中冷眼以待的人现在向他伸出援助之手。在帮助溺水者的同时而又不危及自身，这需要慎之又慎。

NEVER DIE OF ANOTHER'S BAD LUCK.

Notice those who stick in the mud, and observe how they call others to their aid so as to console themselves with a companion in misfortune. They seek someone to help them to bear misfortune, and often those who turned the cold shoulder on them in prosperity now give them a helping hand. There is great caution needed in helping the drowning without endangering oneself.

www.buhua.com
2004

不要对所有人负责，也不要对每个人负责

否则的话你会变成奴隶，而且是所有人的奴隶。有的人生来就比别人更幸运；他们生来就施惠于人，而别人则安而受之。自由，比你可以为之而放弃自由的任何礼物都更加珍贵。与其让许多人依赖于你，不如让你自己独立于任何人。权力的唯一优势，就是让你能够做更多的善事。尤其是不要把责任视为恩惠，因为它通常是别人的计谋，为的是让你依赖于他。

DO NOT BECOME RESPONSIBLE FOR ALL OR FOR EVERYONE.

Otherwise you become a slave and the slave of all. Some are born more fortunate than others; they are born to do good as others are to receive it. Freedom is more precious than any gifts for which you may be tempted to give it up. Lay less stress on making many dependent on you than on keeping yourself independent of any. The sole advantage of power is that you can do more good. Above all do not regard a responsibility as a favor, for generally it is another's plan to make you dependent on him.

不要凭激情行事

如果你这么做，就会失去一切。如果你迷失了自己，就不能代表自己，激情总是放逐理智。这种时候，应该引入一个能保持冷静的审慎的中间人。之所以当局者迷，旁观者清，也就是因为旁观者能保持冷静。一旦觉察到自己的情绪正在失控，你就应该明智地迅速收兵。一旦热血沸腾，很快就会血光四溅。冲动一时，后悔一世；使自己遭殃，招旁人怨怼。

NEVER ACT OUT OF PASSION.

If you do all is lost. You cannot act for yourself if you are not yourself, and passion always drives out reason. In such cases interpose a prudent go-between who can keep cool. That is why onlookers see more of the game, because they keep cool. As soon as you notice that you are losing your temper beat a wise retreat. For no sooner is the blood up than it is spilled. A few moments may be given for many days' repentance for oneself and complaints from others.

顺势而为

我们的思想行为以及所有的一切，都必然被环境所决定。可以做的时候就行动吧，因为时不我待。不要按照一成不变的原则去生活，除了那些关乎基本道德的原则之外。也不要让你的意志成为固定环境的抵押物，因为你明天没准要喝今天泼掉的水。有些人是如此荒谬地自相矛盾，以至于期望一次行动的所有环境都能服从于他们的奇思妙想，而不是相反。智者懂得，审慎的指导原则就在于见风使舵。

LIVE FOR THE MOMENT.

Our acts and thoughts and all must be determined by circumstances. Act when you may, for time and tide wait for no one. Do not live by certain fixed rules, except those that relate to the cardinal virtues. Nor let your will pledge to fixed conditions, for you may have to drink the water tomorrow that you cast away today. There are some so absurdly paradoxical that they expect all the circumstances of an action should bend to their eccentric whims and not vice versa. The wise man knows that the very polestar of prudence lies in steering by the prevailing wind.

咱们也变身吧！
几口
坚决不！

最贬损一个人的，莫过于说他跟任何人都很像

NOTHING DEPRECIATES A PERSON MORE THAN TO SHOW HE IS JUST LIKE ANYONE ELSE.

他看上去太有人味的那一天，也就不再被认为是神了。轻浮跟声望背道而驰。正如矜持者被认为比常人更高一样，轻浮者也被认为比常人更低。导致你不受人尊敬的缺点，莫过于轻浮。因为轻浮跟严肃背道而驰。一个轻浮的人不可能是个有分量的人，哪怕当他已经老了、年龄应该迫使他谨慎的时候。尽管这种缺点如此普遍，但它依然被人蔑视。

The day he is seen to be all too human he ceases to be thought divine. Frivolity is the exact opposite of reputation. And as the reserved are held to be more than men, so the frivolous are held to be less. No failing causes failure of respect. For frivolity is the exact opposite of solid seriousness. A person of levity cannot be a person of weight even when he is old, and age should oblige him to be prudent. Although this blemish is so common it is none the less despised.

观众朋友~明天又有风了啦！好讨厌的！人家不喜欢！
天气预报
电视台
乌鲁木齐
6~18
长春
1~13
北京
重庆
12~20
长沙
10~17
南昌
11~18
贵阳
4~11
昆明
15~23
广州
16~20

既被人爱,又被人敬,是为大幸

IT IS A PIECE OF GOOD FORTUNE TO COMBINE PEOPLE'S LOVE AND RESPECT.

通常,一个人如果想受人尊敬,就不敢被人爱之太殷。爱比恨更脆弱。爱与敬无法携手同行。所以,一个人既不可使人畏之太甚,亦不可使人爱之太深。爱以信任为起点,爱进则敬退,此消则彼长。宁愿别人以尊敬之心爱你,而不要他们以热烈之情爱你,因为敬爱是一种对多数人都适合的爱。

Generally, one dare not be liked if one would be respected. Love is more sensitive than hate. Love and honor do not go well together. So that one should aim neither to be much feared nor much loved. Love introduces confidence, and the further this advances the more respect recedes. Prefer to be loved with respect rather than with passion, for that is a love suitable for many.

www.buhua.com
2004

懂得如何测试人

智者的小心必定指向卑劣者的陷阱。良好的判断力需要测试他人的判断力。了解人的品格和特性，比了解蔬菜与矿物的特性更重要。事实上，这是生活中最微妙的事情之一。听音识金，辨言知人。言辞是诚实的证据，行为则更是如此。在这里，你需要非凡的主意力，深刻的观察力，敏锐的识别力，以及明智的判断力。

KNOW HOW TO TEST PEOPLE.

The care of the wise must guide against the snare of the wicked. Great judgement is needed to test the judgement of another. It is more important to know the characteristics and properties of people than those of vegetables and minerals. Indeed, it is one of the shrewdest things in life. You can tell metals by their ring and men by their voice. Words are proof of integrity, deeds still more. Here one requires extraordinary care, deep observation, subtle discernment, and judicious decision.

让你的个人品质超过你的职责需求

LET YOUR PERSONAL QUALITIES SURPASS THE REQUIREMENTS OF YOUR OFFICE.

要让你的个人品质超过你的职责需求,而不是相反。不管职位多高,人应该更高。当你的职位变得越高的时候,你的才能必须越来越广泛地延展扩充。另一方面,心胸狭窄的人很容易灰心丧气,最终因职责和名声的日渐减损而徒自伤悲。伟大的奥古斯都[1]更看重自己是个伟大的男人,而不是一位伟大的君王。因此,胸怀高远者总能找到用武之地,基础扎实者必能发现天赐良机。

Do not let it be the other way about. However high the post, the person should be higher. An extensive capacity extends and dilates more and more as his office becomes higher. On the other hand, the narrow-minded will easily lose heart and come to grief with diminished responsibilities and reputation. The great Augustus thought more of being a great man than a great prince. Here a lofty mind finds fit place, and well-grounded confidence finds its opportunity.

---

① 奥古斯都(前 63-14),罗马帝国第一任皇帝(前 27-14),恺撒的侄孙。

www.buhua.com
2004

成熟

成熟既表现在穿着打扮上，更表现在行为习惯上。物质的重量是贵金属的标志，精神的重量是贵人的标志。成熟使他的能力得以完美，唤起人们的尊敬。在人的身上，沉着镇静的仪态举止构成了其灵魂的外表。它只存在于平静威严的派头中，蠢人的懵懂无知中没有这个，而只有轻浮。在前者那里，言必信，行必果。成熟让人得以完美，因为迄今为止每个人的完善程度都是取决于他的成熟。在不再是个孩子的时候，一个人就开始获得严肃与威严。

MATURITY.

It is shown in the costume, still more in the customs. Material weight is the sign of a precious metal, moral weight is the sign of a precious man. Maturity gives finish to his capacity and arouses respect. A composed bearing in a person forms a façade to his soul. It does not consist in the insensibility of fools, as frivolity would have it, but in a calm tone of authority. With people of this kind sentences are orations and acts are deeds. Maturity puts a finish on a person for each is so far complete only according as he possesses maturity. On ceasing to be a child a person begins to gain seriousness and authority.

欢迎欢迎！隆重欢迎！

在自己的观点上要温和

每个人都是根据自己的利益来抱持自己的观点，并为这些观点设想了大量的理由。因为就大部分人来说，判断力都不得不让路于个人的倾向。很有可能，两个人抱持完全相反的观点，而每个人都认为自己有理，然而道理总是坚持自己的原则，而且绝不会两面三刀。在这种情形下，审慎之人会小心以对，因为他对对方观点的判断或许会让人对自己的观点产生怀疑。不妨把自己置于对方的位置上，然后研究其观点形成的原因。那么你就不会以这样一种稀里糊涂的方式宣告对方是错的、证明自己是对的。

BE MODERATE IN YOUR VIEWS.

Everyone holds views according to his interest, and imagines he has abundant grounds for them. For with most people judgement has to give way to inclination. It may occur that two may meet with exactly opposite views and yet each thinks to have reason on his side, yet reason is always true to itself and never has two faces. In such a situation a prudent person will proceed with care, for his judgement of his opponent's view may cast doubt on his own. Place yourself in the other person's place and then investigate the reasons for his opinion. You will not then condemn him or justify yourself in such a confusing way.

美好无限

别虚张声势

DO NOT AFFECT WHAT YOU HAVE NOT EFFECTED.

许多人总是毫无理由地宣布自己的成就。他们冷静沉着地制造了一个所有人都搞不懂的神话。他们是欢呼喝彩的变色龙，奉献给他人的是过度的笑声。虚荣总是令人讨厌。那些荣誉的蚂蚁到处爬来爬去，窃取功绩的残羹剩饭。你的功绩越大，就越不需要炫耀。只满足于做，而把说留给别人吧。把你的功绩赠送掉，而不要卖掉。不要用腐败的笔在泥泞中写下赞美的话语，嘲笑那些知道得更清楚的人。要立志做一个英雄，而不仅仅是看上去像个英雄。

Many claim accomplishments without the slightest cause. With great coolness they make a mystery of all. Chameleons of applause they afford others a surfeit of laughter. Vanity is always objectionable, here it is despicable. These ants of honor go crawling about filching scraps of exploits. The greater your exploits the less you need affect them. Content yourself with doing, leave the talking to others. Give away your deeds but do not sell them. And do not hire venal pens to write down praises in the mud, to the derision of those who know better. Aspire rather to be a hero than merely to appear to be one.

全宇宙最牛作品制作中！
严禁入内

高贵的品质

高贵的品质造就高贵的人，一项高贵的品质比一大堆平凡的品质更有价值。曾经有人总是让自己的财产、甚至家用器具尽可能地大。一个伟大的人更应该认识到，要让自己的灵魂尽可能大。在上帝那里，一切都是永恒的、无穷的；在英雄那里，每件事物都应该是伟大的、庄严的，所以，他的一切行为——不，他的一切言辞——都应该充满了超凡出众的威严。

NOBLE QUALITIES.

Noble qualities make noble people; a single one of them is worth more than a multitude of mediocre ones. There was once a man who made all his belongings, even his household utensils, as great as possible. How much more ought a great man see that the qualities of his soul are as great as possible. In God all is eternal and infinite; in a hero everything should be great and majestic, so that all his deeds — no, all his words-should be pervaded by a transcendent majesty.

一举一动，
都仿佛别人在注视着你

你必须知道，周围的人都在看着你，或者将会看到你。你要懂得隔墙有耳，懂得坏事传千里的道理。即使是独自一人，你的一举一动也要仿佛全世界的眼睛在看着你。因为你知道，一切迟早要被人知道，所以你就会把那些后来听闻你的行为的人看作是近在眼前的目击者。希望全世界始终在看着自己的人，当然不在乎隔壁邻居越过高墙看自己。

ALWAYS ACT AS IF OTHERS WERE WATCHING.

He must see all round who sees that men see him or will see him. He knows that walls have ears and that ill deeds rebound back. Even when alone he acts as if the eyes of the whole world were upon him. For he knows that sooner or later all will be known, so he considers those to be present as witnesses who must afterwards hear of the deed. He that wished the whole world might always see him did not mind that his neighbors could see him over their walls.

三样东西造就天才

它们是上天慷慨馈赠的最完美的天赋——丰富多产的创造力，深邃渊博的智力，以及令人愉快而优雅的品味。思考周全固然好，思考正确则更好——这是对善的理解。它不会取代判断力而居于中枢的位置，它更多的属于烦恼，而非效用。思考正确是合理天性之果。20岁时意志为王，30岁时智力做主，40岁时判断力当家。有些人的心智就像山猫的眼睛在暗处闪亮，最黑暗的地方看得最清楚。有些人更适合临场发挥——他们总是灵机一动，突然发现适合于突发事件的应急措施；这种品质能结出更多更好的硕果，是一种多产的福气。与此同时，良好的品味则适合于人的整个一生。

THREE THINGS GO TO A PRODIGY.

They are the choicest gifts of Heaven's prodigality — a fertile genius, a profound intellect, a pleasant and refined taste. To think well is good, to think right is better — it is the understanding of the good. It will not do for the judgement to reside in the backbone; it would be of more trouble than use. To think right is the fruit of reasonable nature. At twenty the will rules, at thirty the intellect, at forty the judgement. There are minds that shine in the dark like the eyes of a lynx, and are most clear where there is most darkness. Others are more adapted for the occasion — they always hit on that which suits the emergency; such a quality produces much and good, a sort of fertile felicity. In the meantime, good taste seasons the whole of life.

人

适当保持饥饿

即便是盛满玉液琼浆的酒碗，你也应该适时从唇边移开。需求乃是价值的尺度。即使是肉体上的饥渴，也只能适当满足，而不是彻底消除，这才是良好品味的标志。好而少，倍加好。梅开二度，其艳必衰。过分的餍足总是危险的，它使最高力量堕于恶意。通过保持饥饿来吊足胃口，是获取满足的不二法门。如果你必须刺激需求，更有效的方法是保持愿望迫切，而不是放任享受过度。努力挣得的幸福，将使快乐倍增。

LEAVE OFF HUNGRY.

One ought to remove even the bowl of nectar from the lips. Demand is the measure of value. Even with regard to bodily thirst it is a mark of good taste to slake but not to quench it. Little and good is twice good. The second time around comes as a great falling off. Too much pleasure is always dangerous and brings down the ill will of the highest powers. The only way to please is to revive the appetite by the hunger that is left. If you must excite desire, better do it by the impatience of want than by the surfeit of enjoyment. Happiness earned gives double joy.

www.buhua.com
2004

嘉言懿行使人完美

一个人应该言辞得体，举止优雅，前者是头脑的杰出，后者是心灵的卓越，二者都来自于灵魂的高贵。言辞是行为的影子，前者为阴，后者为阳。享有声望比传播声望更重要。言甚易，行甚难。言为文，行为质，文质彬彬，然后君子。良行耐久，嘉言易亡。思为行之本，行为思之果；若思想慎明，则行之有效。

WORDS AND DEEDS MAKE THE PERFECT PERSON.

One should speak well and act honorably the one is an excellence of the head, the other of the heart, and both arise from nobility of soul. Words are the shadows of deeds — the former are feminine, the latter masculine. It is more important to be renowned than to convey renown. Speech is easy, action hard. Actions are the stuff of life, words its frippery. Eminent deeds endure, striking words pass away. Actions are the fruit of thought; if this is wise, they are effective.

www.buhua.com
2004

一言以蔽之：做个圣徒

对所有人而言，都是如此。美德，是连接众善的链环，是一切幸福的中心。她使你明智、审慎、睿智、细心、聪敏、勇敢、体贴、忠诚、快乐、可敬、诚实，使你成为一个完美无缺的英雄。有三样东西使人幸福：健康、圣洁和智慧。美德是凡尘俗世的太阳，它的运行轨道就是天地良心。她是如此美轮美奂，既赢得了万能之神的宠幸，也获得了普世众生的青睐。没什么东西有美德可爱，没什么东西比邪恶可恨。衡量一个人的才能和伟大，只能通过他的美德，而非他的运气。唯有美德是完全自足的。她使得我们生前受人爱戴，死后叫人怀念。

IN ONE WORD, BE A SAINT.

So is all said at once. Virtue is the link of all perfections, the center of all the felicities. She makes a person prudent, discreet, sagacious, cautious, wise, courageous, thoughtful, trustworthy, happy, honored, truthful, and a universal hero. Three things make a person happy — health, holiness, and wisdom. Virtue is the sun of our world, and has for its course a good conscience. She is so beautiful that she finds favor with both god and man. Nothing is lovable but virtue, nothing detestable but vice. A person's capacity and greatness are to be measured by his virtue and not by his fortune. She alone is all-sufficient. She makes people lovable in life, memorable after death.